Operator Theory
Advances and Applications
Vol. 72

Editor
I. Gohberg

Matrix and Operator Valued Functions

The Vladimir Petrovich Potapov Memorial Volume

Edited by

I. Gohberg
L.A. Sakhnovich

Springer Basel AG

Volume Editorial Office:

Raymond and Beverly Sackler Faculty of Exact Sciences
School of Mathematical Sciences
Tel Aviv University
69978 Tel Aviv
Israel

A CIP catalogue record for this book is available from the Library of Congress, Washington D.C., USA

Deutsche Bibliothek Cataloging-in-Publication Data
Matrix and operator valued functions : the Vladimir Petrovich
memorial volume / ed. by I. Gohberg ... – Basel ; Boston ;
Berlin : Birkhäuser, 1994
  (Operator theory ; Vol. 72)
  ISBN 978-3-0348-9667-2   ISBN 978-3-0348-8532-4 (eBook)
  DOI 10.1007/978-3-0348-8532-4
NE: Gochberg, Izrail' (Hrsg.) ; Potapov, Vladimir Petrovich : Gedenkschrift ; GT

© 1994 Springer Basel AG
Originally published by Birkhäuser Verlag in 1994
Softcover reprint of the hardcover 1st edition 1994

Printed on acid-free paper produced from chlorine-free pulp
Cover design: Heinz Hiltbrunner, Basel

9 8 7 6 5 4 3 2 1

Table of Contents

EDITORIAL INTRODUCTION

This book is dedicated to the memory of an outstanding mathematician and personality, Vladimir Petrovich Potapov. V. Potapov worked on the border between operator theory and complex analysis, and he made important contributions and had a great deal of influence in both areas.

The first part of the book consists of biographical and personal material on V. Potapov. The rest of the book is a collection of papers on different aspects of operator theory and complex analysis covering the recent achievements of the Odessa-Kharkov school in areas in which Potapov was very active. Some of the papers describe the achievements of V. Potapov and give an account of the latest state of affairs, while others present continuations and generalizations, as well as related problems and results.

The publication of this book was made possible by the efforts of a large number of mathematicians. Three years ago, during the 1991 Toeplitz Lectures held in Tel Aviv University, the editors of this volume took the initiative and made plans for this book. The next step was to invite papers from people who were students, colleagues or followers of V. Potapov. The papers were then sent to the editorial office. At this stage they were in the form of rough translations, containing handwritten formulas. A Western team of experts consisting of D. Alpay, J.A. Ball, A. Dijksma, H. Dym, A. Iacob, M.A. Kaashoek, H. Landau, A.C.M. Ran, L. Rodman, and J.L. Rovnyak, then helped the editorial office to process the papers, in particular to do the copy editing and retyping of the manuscripts. The manuscripts then had to be sent to the authors for proofreading, with the subsequent mailing difficulties encountered. We often overcame the above difficulties by using special messengers to deliver the papers to the authors and return them after proofreading to the editorial office for further processing. All of this was no mean task. With the further help of the above team the manuscripts were then put into their final camera-ready form. The photographs were kindly offered to us by V. Katsnelson.

We would like to use this opportunity to thank all colleagues involved for their help in producing this book containing an interesting selection of papers related to V.P. Potapov. We believe it will be of interest to a large audience, and will pay tribute to the memory of an outstanding colleague, teacher and friend.

VLADIMIR PETROVICH POTAPOV

VLADIMIR PETROVICH POTAPOV

Vladimir Petrovich Potapov was born on January 24th, 1914. He was the youngest son in the Potapov family. His father, Petr Osipovich, a privat-docent at Novorossiysky University in Odessa, was the son of a landless peasant from Tambov province. At the age of five Petr Osipovich began to work in the fields. Nevertheless, thanks to his native wit and excellent abilities, he graduated from the Nezhin Gymnasia in 1902 with a gold medal and was taken on the staff of the Nezhin Historical and Philological Institute. At the gymnasia he met Anna Mihailovna Novicka, whom he later married. In 1911 P.O. Potapov began to lecture on Old Russian literature at the University in Odessa.

V.P. Potapov was thus born into the family of a highly educated philologist, and a connoisseur of Slavic languages and literature. He did not attend school. On leaving the gymnasia Anna Mihailovna worked for some years as a teacher and she came to the conclusion that school stultifies. When Vladimir was six, P.O. Potapov began to teach him mathematics, history, literature and languages. Music ranked high in the Potapov household. A special teacher was engaged to give Vladimir music lessons. These were supplemented by his mother, Anna Mihailovna, who also taught him literature. His father, Petr Osipovich, also had an ear for music. He had a good voice and could play the guitar.

In 1931, Vladimir Petrovich entered the Odessa conservatory. He was well educated and, although only seventeen, very serious. Whatever he did, he did in a very serious fashion; he walked and spoke slowly, and had his own opinion about everything. Even his teachers at the conservatory respected his profundity and thoroughness and called him "professor". And, as time proved, they were not mistaken.

At the Odessa conservatory, Vladimir made friends with Emil Gilels and Ludmila Ginzburg. The friendship of these three so different young people lasted for life. Emil Gilels, an outstanding musician in his own right, always took Potapov's opinion into consideration, and highly valued his advice. V.P. Potapov was no ordinary music connoisseur, but a musician with his own ideas, and he felt it necessary to share them with others. His daughter Tatjana was born in 1949. In the early 60's he began to give music lessons to Tatjana's classmate, whom he recognized as a very gifted boy. It is certainly unusual that a middle-aged professor of mathematics should visit a youngster every day to teach him music, no matter how talented the boy is. Tuition fees were out of the question. V.P. was not mistaken in his student; Boris Blokh, now a professor at the Essen conservatory, developed into a world-famous pianist.

L.N. Ginzburg, a professor at the Odessa conservatory, recollects: "Vladimir was much more educated than any of us and could speak about music very vividly and figuratively. In those years he admired Prokofiev. In his free time he often he played Prokofiev's cycle "Transiencies", and we watched and listened to him. But, in spite of his deep understanding of music, he did not have the natural agility required to become a great pianist. Gilels, who was a great mimic, often made us laugh by imitating Vladimir playing".

V.P. did not enter the fourth year at the conservatory. One of Potapov's distinguishing traits was his ability to weigh the pros and cons and to come to a decision, which was at times unexpected. He gave up his studies in music and took the examinations for admission to the Faculty of Physics and Mathematics of Odessa University. He passed these examinations brilliantly and was accepted as a student.

At that time the now famous Odessa school of mathematics sprung up around Mark Grigorevich Krein. The younger participants included M.S. Livšic, F.A. Naimark, I. M. Glazman, F.A. Rutman, D.P. Milman, M.S. Brodsky, S.A. Orlov and L.S. Dašnic, all of whom studied at the university together with Potapov.

In 1939, V.P. became the Ph.D. student of Professor B.Ya. Levin. Levin's charm and scientific interests determined Potapov's choice. He decided to focus on the problem of the divisor's of almost periodic polynomials. V.P. also worked as an assistant lecturer at the university and he combined the writing of his thesis with his teaching activities. From 1940 until the outbreak of the Great Patriotic War on June 22, 1941 he worked at the Odessa Institute of Marine Engineering. He was exempted from military service because of severe myopia.

In 1941 he and his wife A.E. Borzon, a pianist whom he met at the conservatory and married in 1939, were evacuated to the village of Fari, where he worked as a teacher of mathematics. A year later he moved to Yerevan. V.P. Potapov worked there as a teacher at the secondary school, and at the same time as a lecturer at the Russian Teacher's Training Institute until 1944.

In 1944 he received an offer from the Odessa Institute of Marine Engineering, and returned to Odessa, where he worked as an assistant lecturer in the Department of Mathematics for one year. All these years, in spite of his great teaching load, V.P. continued to work on his Ph.D. thesis. In 1945, at the Institute of Mathematics of the Academy of Sciences of the USSR, V.P. Potapov defended his first Ph.D. thesis. His advisor, B.Ya. Levin always rated it higher than P.V.'s subsequent fundamental works.

Potapov always carried his work through to the end. It was one of the main traits of his character. His work absorbed him completely. He always gave the problems he was working on a new and deeper meaning, which revealed their inner logic. He could work on a theme for several years with unremitting persistence, and could not be hurried to publish the results by a fixed date.

After the war, M.G. Krein, B.Ya. Levin and their students were not reinstated at the Odessa University. Because of this, the scientific and pedagogical standards of the mathematics department were considerably lowered. Students refused to attend the lectures of some professors. In September 1946, *The Bolshevic Banner* (an Odessa newspaper) carried an article by V.P. Potapov and the journalist D. List. The authors of this article described in detail the disastrous state of the Faculty of Physics and Mathematics. They ended the article with the words: "The leaders of the university have to correct their mistakes and to revive immediately the famous traditions of the Odessa School of Mathematics. The Faculty of Physics and Mathematics has to become active again, as a really creative center of scientific and mathematical thought in our city".

After that the Rector of the university was obliged to invite M.G. Krein, B.Ya. Levin, and M.A. Rutman to return. However, in April of 1948 M.G. Krein and B.Ya. Levin were dismissed. Soon an article appeared in which M.G. Krein and B.Ya. Levin were proclaimed to be "rootless cosmopolitans". At that time many Jewish scientists, writers, and other cultural leaders were exposed to public dishonour by these words. During the forties and fifties V.P. Potapov made many unsuccessful attempts to bring M.G. Krein and his students back to the university. The leaders of the Faculty of Physics and Mathematics could not forgive him for this. In fact they prohibited the use of any subject connected with Potapov's investigations as a theme for a diploma.

In 1945 V.P.'s father, Professor P.O. Potapov died. From 1932 he had been the Dean of the Faculty of Language and Literature and had headed the Department of Russian Language at the Pedagogical Institute. He also headed the Department of Russian Language and Slavic Philology at the Odessa University since 1937. Petr Osipovich was fondly remembered. A scientific conference dedicated to the centenary of his birth was held in Odessa 1989.

In 1948 B.Ya. Levin, who held a position at the same Pedagogical Institute, invited Vladimir Petrovich to work there. Soon V.P. became the head of the Department of Mathematics, and from 1952 held the position of Dean of the Faculty of Physics and Mathematics. V.P. managed to enlist to the faculty most of the scientists of Krein's school: M.S. Livšic, M.A. Rutman, D.P. Milman, S.A. Orlov, D.I. Kucher, and L.S. Dašnic all lectured at the Institute. As noted earlier, M.G. Krein and other Jewish lecturers were dismissed from Odessa University, and most of the really good Odessa mathematicians worked in other academic institutions. In 1944 M.S. Livšic defended his second Ph.D. thesis and was working out the theory of nonselfadjoint operators, which was generally recognized to be of fundamental importance by mathematicians the world over. The concept of a characteristic matrix function was the basic corner stone of the theory. The analysis of matrix functions became the focal point of Potapov's research. In the fifties V.P. worked out the theory of J-contractive matrix functions. In 1954 he defended his second thesis at the Institute of Mathematics of the Academy of Sciences of the USSR. In 1955 it was published in *Trudy Moscovskogo Mathematicheskogo Obschestva*. This work laid down the foundations of the J-theory.

Potapov's theory gave a new approach to solving problems in the theory of interpolation, and the theory of electrical networks. During his last years he continued to develop and popularize the J-theory. He planned to write a monograph on that theory and its applications, but the plan was never realized. After his death others (H. Dym and V.E. Katsnelson) published monographs on the J-theory.

M.S. Livšic ran a mathematical seminar which was the epicentre of the scientific life of the faculty. It was attended by lecturers of other institutes, and V.P. also invited a number of students to participate. In 1950 a gifted second year student, L.A. Sakhnovich, caught his attention, and he began to tutor him. V.P. recognized this student's abilities and set him to solve a rather difficult problem. Sakhnovich solved the problem, and this marked the beginning of his scientific career.

Potapov's excellent qualities were illustrated most sharply in personal contacts with his students. He not only appreciated cleverness and ability in others highly, but he also felt a persistent need to help them. V.P. was a teacher, an advisor, and a friend to each of his students. One of his students, who later became a professor, said that remembering Vladimir Petrovich helped him not to give way to anger and irritation. Even when V.P. left Odessa he kept himself informed about the scientific and private life of his students. He was proud of their scientific achievements. Nevertheless, because of his very high standards, it was difficult to get his enthusiastic approval. He was exacting and quite often criticized the work of his colleagues and students.

In conversation with his students, V.P. displayed a high intellectual and cultural level, and he liked to test them when an opportunity presented itself.

"Do you know who is the prototype of Jean Christophe?" he would ask unexpectedly.

"What do you tell your students about Tutchev?" When L.A. Sakhnovich answered him that the changes of rhythm in Tutchev's poems coincided with the changes in the rhythm of the heartbeat he nodded approval: "Such lectures can be given".

D.Z. Arov, Yu.P. Ginzburg, I.V. Kovalishina, L.B. Golinsky, V.K. Dubovoy, A.V. Efimov, T.S. Ivanchenko, E.Ya. Melamud, L.A. Sakhnovich and I.V. Mikhailova, were all students of V.P.

Every phrase of V.P.'s lectures was well thought out and complete, even the formulae were calligraphically written out. The lecture was not given but "drawn" on the blackboard, and V.P.'s commentary was unhurried and deep. His speech was picturesque, and sometimes sharp. Once he was giving a lecture on applications of the J-theory to the theory of electrical networks at the Odessa Electrical Engineering Institute. He felt that the listeners did not understand the material and concluded the lecture with these words: "To teach practical workers theory is about the same as to milk a bull". His listeners were quick to agree with him. V.P. paused and added: "But everybody knows that a bull has other functions". We can mention another episode. A certain N., who was a lecturer in Marxism, handed in a report on the politically weak students. The report ended with a demand "to take prompt action". Potapov looked it through, underlined 17 grammatical errors, and gave it to the university administration with a recommendation "to take prompt action". Measures were taken and N. was transferred to another faculty.

The wealth of historical, literary and biblical associations which were part and parcel of Potapov's everyday speech made conversation with him even more profound and many-sided. His arguments were exact and he was not afraid to mince words.

On one occasion he argued in favor of approving someone's first Ph.D. thesis on the grounds that: "His thesis clearly shows that this problem is not worth solving".

In 1954 V.P. became a member of the Communist Party. As he said, "it will be nice to have some respectable people in the party". In contrast to many others, V.P. managed to remain an honest person, even within the ranks of the party.

He had no illusions about "the historical role of the party". His colleagues knew his statement "I would gladly be hanged if the whole party is hanged together with me". Membership in the party gave him an opportunity to take an active civic stand to help his gifted students to be accepted into the Ph.D. program and to obtain employment for them afterwards.

In 1956 V.P. was awarded the chair of mathematics at the Odessa Technological Institute of Food and Refrigerating Industry, but up until 1963 he continued to lecture at the Pedagogical Institute. A son Andrei was born in 1962.

At that time he found new ways of developing the J-theory, and its connections with problems of interpolation and network theory. P.V.'s scientific style was determined by his wholehearted nature. He felt very strongly that a problem had to be solved by the methods of the theory in the terms in which it had been formulated. He maintained the idea that the methods of operator theory which were often used were besides the point. He formulated this idea succinctly and picturesquely: "Operator theory is a parasite on the clean body of function theory".

In spite of the fact that Potapov's work had significant applied importance he never did industrial consulting to supplement his income. He did not want to be distracted from the chief orientation of his research. Once, when talking to a colleague, V.P. said with the laconicism and figurativeness so characteristic of him: "To do scientific work within the bounds of an outside-contract is just the same as to send your wife to walk the streets". During those years Professor Martynovsky was the Rector of the Technological Institute. He was an eminent scientist and understood the importance of mathematics and V.P.'s place in it. After his death, V.P.'s situation at the Institute deteriorated. The government education policy encouraged institutes to accept the children of workers and peasants even if they were not qualified to be students. The new administration accused V.P. of giving them bad marks. Earlier he had also been blamed for distributing a large number of low grades. In 1953 during a meeting of the academic council at the Pedagogical Institute, V.P. said: "After all, a bad mark in mathematics is no obstacle to a subsequent career. I was recently in the town of Gory, the birthplace of the leader of the people, J.V. Stalin, and I saw a bad mark in mathematics in his school progress record". A lengthy silence followed this remark, and the council broke up without the members saying a word to each other. V.P. was not subjected to repression, which can be explained by both the decency of the members of the council and the death of Stalin soon afterwards. However, times had changed. V.P. was forced to leave the Technological Institute. From 1974 he lectured at the Odessa Institute of National Economy.

In 1976 V.P. left for Kharkov to head the Department of Applied Mathematics at the Institute for Low Temperature Physics and Engineering. As in Odessa, he was surrounded by students and followers.

V.P. organized his department with great energy, but more and more often he thought about mortality. In 1975, when speaking with friends, he repeated over and over again that he had only five more years to live. He enthusiastically popularized the J-theory. He lectured on it in Odessa and Erevan, and organized his own school within the framework of the mathematical Winter School in Voronezh. He thought of writing a monograph on the J-theory, but that was not to be.

Vladimir Petrovich died on December 21, 1980.

D.Z. Arov
Yu.P. Ginzburg
S.M. Pozin
L.A. Sakhnovich

Translated by I.E. Shtein

V. P. POTAPOV, WINTER 1977

V.P. POTAPOV

CURRICULUM VITAE

Date of birth: January, 24, 1914
Date of death: December, 21, 1980

Education:

M.Sc. Mathematics
Odessa University, Odessa, Ukraine
Date of award: 1939

Kandidat of Sciences, Ph.D. Mathematics
The Institute of Mathematics of the Academy of Sciences of the USSR
Date of award: 1945

Doctor of Sciences, Mathematics
The Institute of Mathematics of the Academy of Sciences of the USSR
Date of award: 1954

Academic and Professional Experience:

1939-1940 Assistant Professor, Odessa University

1940-1941 Assistant Professor, Odessa Institute of Marine Engineers

1941-1942 Teacher of Mathematics,
Secondary School Fari, Northern Osetia

1942-1944 Teacher of Mathematics,
Secondary school, Assistant Professor,
Jerevan Russian Teachers' Institute, Jerevan, Armenia

1944-1945 Assistant Professor,
Odessa Institute of Marine Engineers

1948-1963 Professor,
Chairman of the Department of Mathematics,
Dean of the Faculty of Physics and Mathematics,
Odessa Pedagogical Institute

1956-1974 Chairman of the Department of Mathematics,
Odessa Technological Institute of Food and Refrigerating Industry

1974-1976 Professor,
Odessa Institute of National Economics

1976-1980 Head of the Department of Applied Mathematics,
Physical and Technical Institute of Low Temperatures,
Kharkov, Ukraine.

LIST OF PUBLICATIONS of V. P. POTAPOV

[1]   (with List, D.) *About the Faculty of Physics and Mathematics of Odessa State University*, Bol'shev. Znamia, September 29, 1946. (Russian).

[2]   *On the divisors of an almost-periodic polynomial*, Sbornik. Trud. Inst. Mat. Akad. Nauk Ukrain. SSR (1949), No. 12, 36-81. (Russian).

[3]   (with Livsic, M. S.) *A theorem on the multiplication of characteristic matrix functions*, Dokl. Akad. Nauk SSSR **72** (1950), No. 4, 625-628. (Russian). [MR **11**, 669]; English transl. in Ref. [31], pp. 1-6.

[4]   *On holomorphic matrix functions bounded in the unit circle*, Dokl. Akad. Nauk SSSR **72** (1950), No. 5, 849-852. (Russian). [MR **13**, 736]; English transl. in Ref. [31], pp. 7-12.

[5]   *The multiplicative structure of analytic J-expansive matrix-functions*, Abstract of Doctoral Thesis, V. A. Steklov Mathematics Institute of the Academy of Sciences of USSR, Moscow, 1954. (Russian). [Rzh. Mat. 1955, 2185Д]

[6]   *The multiplicative structure of J-contractive matrix-functions*, Trudy Mosk. Mat. Obshch. **4** (1955), 125-236. (Russian). [MR **17**, 958]; English transl.: Amer. Math. Soc. Transl. (2) **15** (1960), 131-243. [see MR **22** #5733]

[7]   *The multiplicative structure of J-contractive matrix-functions*, Uspekhi Mat. Nauk **10** (1955), No. 3, 185-186. (Russian). [Rzh. Mat. 1958, 289]

[8]   *Analytic matrix-functions*, in: Proccedings of the 3rd All-Union Math. Congress, Moscow, 1956, Vol. 2, p. 34, Akad. Nauk SSSR. (Russian). [Rzh. Mat. 1957, 2251]

[9]   (with Iokhvidov, I. S.) *A poet at heart*, Znamya Kommunizma, December 21, 1962. (Russian). [Rzh. Mat. 1963:5 A29]

[10]   *Application of matrix-functions in the theory of electrical circuits*, in: Proccedings of the 4th All-Union Math. Congress, Leningrad, 1961, Vol. 2, p. 325, Nauka, 1964. (Russian). [Rzh. Mat. 1964:7 Б180]

[11]   (with Kovalishina, I. V.) *Multiplicative structure of analytic real J-dilative matrix-functions*, Izv. Akad. Nauk Armyan. SSR Ser. Fiz.-Mat. Nauk **18** (1965), No. 6, 3-10. (Russian). [MR **35** #5629]; English transl. in Ref. [31], pp. 13-21, and in Ref. [32], pp. 1-8.

[12]   *Multiplicative representations of analytic matrix-functions,* in: Abstracts of Math.
       Section, pp. 74-75, Moscow, August 16-26, 1966. (Russian).

[13]   (with Akhiezer, N.I., Berezanskiĭ, Yu. M., Iokhvidov, I. S., and Mitropol'iĭ, Yu. A.)
       Mark Grigor'evich Kreĭn, Ukrain. Mat. Zh. **19** (1967), No. 5, 114-118. (Russian).
       [MR **35** #6512]

[14]   *General theorems of structure and the splitting of elementary multipliers of analytic
       matrix-functions,* Akad. Nauk Armyan. SSR Doklady **48** (1969), No. 5, 257-263.
       (Russian). [MR **57** # 16627]; English transl. in Ref. [31], pp. 23-31; English transl.
       in Ref. [32], pp. 9-14.

[15]   (with Tovmasyan, T. A.) *A multidimensional gyrator,* Akad. Nauk Armyan. SSR
       Doklady **51** (1969) No. 5, 266-272. (Russian). [Rzh. Mat. 1971:6 B406]

[16]   *The main facts of the theory of J-contractive matrix-functions,* in: Abstracts of the
       All-Union Conference on the Theory of Complex Functions, pp. 179-181, Khar'khov
       Physics and Technology Institute for Low Temperatures, Khar'khov, 1971. (in Rus-
       sian).

[17]   (with Efimov, A. V.) *J-expansive matrix-functions and their role in the theory of
       electrical circuits,* Uspekhi Mat. Nauk **28** (1973), No. 1(169), 65-130. (Russian). [MR
       52 #15090]; English transl.: Russian Math. Surveys **28** (1973), No. 1, 69-140.

[18]   (with Kovalishina, I. V.) *An indefinite metric in the Nevanlinna-Pick problem,* Akad.
       Nauk Armyan. SSR Doklady **59** (1974), No. 1, 17-22. (Russian). [MR **53** #13577];
       English transl. in Ref. [31], pp. 33-40, and in Ref. [32], pp. 15-20.

[19]   *Linear-fractional transformations of matrices,* in: Studies in the Theory of Operators
       and their Applications, pp. 75-97, Naukova Dumka, Kiev, 1979. (Russian). [MR
       81f:15023]; English transl.: in Ref. [31], pp. 75-97, and in Ref. [32], pp. 21-36.

[20]   (with Mikhaĭlova, I. V.) *On a criterion of Hermitian positivity,* Physics and Tech-
       nology Institute for Low Temperatures, Kiev, 1980, Preprint No. 17. (in Russian).

[21]   (with Kovalishina, I. V.) *Radii of the Weyl circle in the matrix Nevanlinna-Pick
       problem,* in: Theory of Operators in Function Spaces and its Applications, pp. 25-
       49, Naukova Dumka, Kiev, 1981. (Russian). [MR **84j:30050**]; English transl. in Ref.
       [31], pp. 25-49, and in Ref. [32], pp. 37-54.

[22]   (with Mikhaĭlova, I. V.) *A criterion for Hermitian positivity,* Dokl. Akad. Nauk
       Ukrain. SSR Ser. A, No. 9 (1981), 22-27. (Russian). [MR **83d:47030**]

[23]   (with Mikhaĭlova, I. V.) *On a criterion of Hermitian positivity,* Teor. Funktsiĭ
       Funkstional. Anal. i Prilozhen., No. 36 (1981), 65-89. (Russian). [MR **84a:47051**]

[24]   (with Kovalishina, I. V.) *Integral representation of Hermitian positive functions,* Khark'ov Railway Engineering Inst., Khar'kov, 1981. Deposited in VINITI 6.19.1981, No. 2984-81. (Russian); English transl.: *Integral representation of Hermitian positive functions,* Translated from the Russsian by T. Ando, Hokkaido University, Sapporo, 1982. [MR **85e:47019**]

[25]   *On the theory of the Weyl matrix circles,* in: Functional Analysis and Applied Mathematics, pp. 131-121, Kiev, 1982. (Russian). [Rzh. Mat. 1983:8 Б212]

[26]   *A theorem on the modulus. I. Fundamental concepts. The modulus,* Teor. Funktsiĭ Funkstional. Anal. i Prilozhen., No. 38 (1982), 91-101. (Russian). [MR **84d:15026**]; English transl. in Ref. [32], pp. 55-66.

[27]   *Infinite products and the multiplicative integral,* in: Analysis in Infinite-Dimensional Spaces and Operator Theory, pp. 117-133, Kiev, 1983. (Russian). [Rzh. Mat. 1984:8 Б1144]

[28]   *A theorem on the modulus. II,* Teor. Funktsiĭ Funkstional. Anal. i Prilozhen., No. 39 (1983), 95-106. (Russian). [MR **85e:47054**]; English transl. in Ref. [32], pp. 67-74.

[29]   (with Kovalishina, I. V.) *The triad method in the theory of extension of a linear function and of an exponential,* Khar'kov Railway Engineering Inst., Khar'kov, 1984. Deposited in VINITI 7.02.1984, No. 4575-84. (Russian). [Rzh. Mat. 1984:10 Б151 Dep.]

[30]   (with Kovalishina, I. V.) *The triad method in the theory of extension of Hermitian positive functions,* Izv. Akad. Nauk Armyan. SSR Ser. Mat. **24** (1989), No. 3, 269-292. (Russian). [MR **91i:42012**]

[31]   *Collected Papers of V. P. Potapov,* Translated from the Russian and edited by T. Ando, Hokkaido University, Research Institute of Applied Electricity, Division of Applied Mathematics, Sapporo, 1982. [MR **84a:01061**]

[32]   (with Kovalishina, I.V.) *Seven papers translated from the Russian,* AMS Translations. Ser. 2. **138** (1988), 1-74.

ODESSA REMINISCENCES

I remember how during one of my visits to Odessa in the mid fifties M.G. Krein and myself were out walking when we met V.P. Potapov with his wife. Mark Gregorievich introduced us. I already knew about V.P. Potapov and his achievements from Krein who considered him to be one of his most brilliant students of whom he was proud. Krein told me about Potapov's second doctoral thesis which he had defended not long before in the Steklov Institute in Moscow. The thesis was about multiplicative decomposition of contractive matrix functions and he considered the results to be fundamental, innovative and opening a new area of research. At that time Potapov was the Dean of the Faculty of Mathematics and Physics of the Odessa Pedagogical Institute and M.G. spoke of him with great respect.

The problem of multiplicative decompositions of contractive matrix functions came from the theory of nonselfadjoint operators and was stated by M. Livsic (a colleague and friend of V.P.). Later M. Livsic used the results of Potapov in the first version of his theory of characteristic functions, the analysis of invariant subspaces of nonselfadjoint operators and their triangular forms. Today these results and their far reaching generalizations are obtained by pure operator theoretic methods. In general Potapov's research very often started with problems appearing in operator theory. Operator theory was used also to predict the results and the formulas for the solutions. At the same time Potapov would usually present his results as pure complex analysis and without operator theory in the final versions of his work. Operator theory had an essential influence on Potapov's work. In its turn his work influenced important research in operator theory and its applications.

In the late forties a wave of antisemitism swept through the Soviet Union and especially Ukraine. M.G. Krein was accused of Jewish nationalism. As usual in such cases no official accusation actually was made, but there were obvious signs that such an accusation was included in the secret files kept in the offices of the Communist Party in Kiev. This accusation was based on the fact that among the first generation of M.G. Krein's Ph.D. students (A.B. Artemenko, M.S. Livsic, D.P. Milman, M.A. Naimark, V.P. Potapov, M.A. Rutman, V.L. Shmulyan) the majority were Jewish. The passage of time proved that M.G. had much more objective reasons for his choice of students. All of them made outstanding contributions to mathematics, but this was of no importance to the party headquarters. These files were often responsible for the fact that M.G. was rejected many times as a candidate for prizes or promotions proposed by most prestigious mathematical organizations. More than that, these files would be used at any time to harm him and his followers, and the outcome could even be a jail or gulag sentence. During these years one of M.G. Krein's students defended his doctoral thesis in Kharkov. There he was officially asked if Krein was still going free.

Among the first generation of M.G. Krein's students only A.B. Artemenko and V.P. Potapov were not Jewish. The former disappeared after the end of World War II, so only Potapov remained. His presence was very important and those in M.G.'s circle understood this very well. I am sure that Potapov also realized the responsibility this carried. Starting with the late forties Krein was forbidden to take on any Ph.D. students who were Jewish. This type of restriction existed in general but V.P. Potapov fought it and was able to overcome it in many cases. D.Z. Arov, L.A. Saknovich and Yu.P. Ginzburg were all Ph.D. students of Potapov during these most difficult years of antisemitism in the Soviet Union. More than that he was very protective of them. Probably no Jew could do this. I can remember the following nontrivial case.

In 1958 an All Union Conference on Functional Analysis was held in Odessa. M.G. Krein was the chairman of the organizing committee, of which V. Potapov was a member. The conference was a success and many outstanding experts from different cities and republics participated. During the organization it turned out that the number of talks to be given by Jews was very high. Since this was an extremely dangerous situation, which could lead to serious consequences for Krein, he suggested imposing some restrictions, in particular not to include in the programme a talk to be given by a young man, Yu. P. Ginzburg, at that time one of V. Potapov's doctoral students. M.G. thought this would not harm the young man and there would be other opportunities for him to give his talk. When Potapov learned about this intention he was furious and threatened to boycott the conference if the talk of Yu. Ginzburg was not included. Ginzburg's talk was included and in this instance everything went smoothly.

I came to know Potapov better some time later when I was trying to arrange for him to move to Kishinev to work in the Moldavian Academy of Sciences. During these days I understood his importance, both as a mathematician and as a personality. I knew he was interested in making this move because at that time he was going through a major change in his personal life (the first of many such changes), and he wanted to move to Kishinev with his new wife. They visited Kishinev and discussed the matter with the authorities of the Academy. I was quite optimistic. To my regret this did not work out and the invitation fell through. Till today the reasons for this are not clear. I can only speculate.

Let me finish with a little anecdote. A colleague was visiting Potapov one day. Potapov offered his visitor a drink, but the latter declined, saying that he had just recently had some heart trouble. Potapov insisted: "The drink will be very good for you. It widens the blood vessels and helps the heart." The colleague replied that after you stop drinking the blood vessels would shrink again and make things even worse. "You are a clever man, but you come to the wrong conclusions. Is it not clear what you have to do - just don't stop." This was Vladimir Petrovich Potapov.

I. Gohberg

THE LAST DAYS OF VLADIMIR PETROVICH POTAPOV

I worked with Vladimir Petrovich and I got to know him very well during the last four years of his life. As opposed to most of his colleagues, chairmen of the departments in our institute, V. Potapov would to come to work every day. However, during the last years it was not easy for him to do so. I used to wait for his arrival with some trepidation. I cannot say that I was afraid of him, but I was somewhat in awe of him, especially at the beginning of our acquaintance.

Our room was at end of a long corridor, and I heard (or rather felt) Vladimir Petrovich ascending the stairs. He climbed slowly, dragging one foot slightly. His laboured breathing betrayed the fact that he was a heavy smoker. Very slowly he would open the door and enter the room. Apart from the three tables which furnished our room (one each for Vladimir Potapov, Ira Michailova and myself) a long row of very uncomfortable creaking old wooden armchairs was ranged along the wall. Potapov would slowly cross the room and with some difficulty he would wedge himself into one of these armchairs. Sometimes he would gaze at the clean blackboard, sucking at his empty cigarette holder. "How are you?" was his usual question. And the working day would start.

In the early days Potapov rented a room on the street where I lived, just five minutes from my house. When I first popped into his house I, as an amateur musician was surprised to see on the wall a colour photograph of Emil Gilels, carrying a signed dedication. (Probably a picture of Cauchy or Riemann would have surprised me much less.) Later I learned that Potapov had a well-deserved reputation in the world of music. He wrote papers for journals and was a close friend of E. Gilels (they were both from Odessa). One time I met V.P. in the narrow corridor of the old (now destroyed) building housing the Philharmonic. It was at a concert of Dimitri Bashkirov. A visit by this world renowned

musician to provincial Kharkov was always a great event, and there wasn't an empty seat
in the hall. I listened to the pianist spellbound. "How do you like it?" I asked V.P. I
considered the question to be rhetorical, but V.P. grimaced and answered "The left hand
was poor." (Maybe he used a more derogatory expression.) I had always considered myself
to have a good musical ear, and I heard nothing wrong. My wife, a professional musician,
also heard nothing wrong.

I was always surprised by his very small, even handwriting. Somehow it did not fit
in with his shapeless figure. Till today I still have his notes on subharmonic functions - a
school notebook written in very small letters containing many remarks made by him with
ballpoint pens of different colours. On the right-hand side of the double spread is the text,
and on the left the notes. How beautifully he wrote (or rather drew) the Gothic letters!
Latterly V.Potapov liked to write on long rolls of used paper from the computer which he
would glue on by himself. He wrote slowly with the minimal number of crossings out or
corrections. He would set his glasses aside and put his head very close to the table. V.P.
hardly ever used the standard abbreviations. He would always write out all the words in
full, even words which were repeated many times. To read, or to decipher, his manuscripts
was very easy and pleasant. However, the roll of paper had to be spread out over the table,
and sometimes even over more than one table.

Sunday, 21st December 1980, I was on duty at the Institute and after work I planned
to visit Potapov who was living in the neighbourhood of the Institute. At that time he
had not been feeling well and did not come to work. The word "sick" somehow did not
apply to him. I am not sure that he had ever been sick at any time in the usual sense of
the word. I was eager to tell him something interesting in connection with the problem on
which I was working, and so I was looking forward to this meeting. I telephoned him in
the afternoon. "Come now!" was his answer.

He was alone at home. Probably he had to make a great effort to come to open the
door, and he needed my help to make his way back. The house had an atmosphere of
loneliness, lack of care and even neglect. However, even at the best of times you could
not call his room tidy. An uncovered sofa, dirty dishes on the table, empty bottles strewn
throughout the room, made a very depressing impression. I cannot say for sure whether

he recognized me. For a time he lay quietly and then his condition worsened drastically. I was at a loss. I started running about the room, looking for medicines, picking up the telephone. "Don't call the ambulance, call Victor" (Katsnelson) were his last words.

I think that there is nothing in the world more terrible than dying in loneliness. Vladimir Petrovich managed to avoid this - and this was all I could do for him.

The laying in state of his body was held two days later in the Institute. It turned out that he had many friends, acquaintances, and colleagues. Among those who spoke were people who had known V.Potapov for many decades. There were the usual long, standard speeches. I was the last to speak. I spoke in the name of our group which consisted of three people, one of whom was Vladimir Petrovich. I can't remember what I spoke about; I think it was about Shostakovich, Hermitian positive functions, and the last work of V.Potapov which he did not manage to complete.

L.B. Golinskii

Operator Theory:
Advances and Applications, Vol. 72
© 1994 Birkhäuser Verlag Basel

THE INFLUENCE OF V. P. POTAPOV AND M. G. KREĬN ON MY SCIENTIFIC WORK

D. Z. Arov

1. MY FIRST DISSERTATION

I first came to know Vladimir Petrovich Potapov and became his graduate student in 1959.

At that time I was as far from his scientific interests as he was from mine. In Odessa, it would have been more natural for me to become a graduate student of M. G. Kreĭn, but at that time this was impossible because for a long time M. G. Kreĭn could not accept Jews as graduate students. V. P. Potapov succeeded in resisting anti-Semitism: before me, Y. P. Ginsburg was his student, and L. A. Sahnovic and E. Y. Melamud remained as graduate students in the Physical-Mathematical Department of the Odessa Pedagogical Institute, where V. P. Potapov was chairman from 1952-56. In later years I heard him say as a joke that Israel would erect a monument for his services to the Jews. V. P. Potapov expected that there would be difficulties in my admission as a graduate student, and he proposed that I ask A. N. Kolmogorov for an assessment of my scientific work. It is possible that this assessment had significance. He also proposed that I first try to become a graduate student at Odessa University under the direction of A. A. Bobrov, who guided my MSc thesis in 1956-57; I wrote two works on probability theory with him. I did attempt this and I was admitted, but only on a correspondence basis. Before my admission, Jews had not been admitted as graduate students since M. G. Kreĭn left the University in 1948. V. P. Potapov suggested that I refuse the "victory" and instead become his graduate student at the Odessa Pedagogical Institute. He applied all possible means to accomplish this.

And now I must write about my "entropy epopee," which explains why V. P. Potapov turned his attention to me, why A. N. Kolmogorov gave me a positive evaluation, and why my first dissertation was executed on a theme so far from Potapov's scientific interests.

In my thesis, which was on information theory, I proposed to apply the notion of entropy to characterize the degree of mixing of dynamical systems. I introduced the entropy $h(T; \xi)$ of a mapping T on a space X with invariant measure μ for a partition ξ of X into measurable sets A_i, and the ϵ-entropy

$$h_\epsilon(T) = \sup\{h(T; \xi) : \xi \text{ with } \mu A_i \geq \epsilon\}.$$

I did not consider

$$h(T) = \sup\{h(T; \xi) : \xi\}$$

because I thought it possible that $h(T) = \infty$ for all T. Since A. A. Bobrov did not work on related problems, he proposed that I go to Moscow University to consult with A. N. Kolmogorov. However, at that time A. N. Kolmogorov was very busy and could only speak with me for about a half hour. He introduced me to R. L. Dobrushin and V. M. Alekseev. To the latter, my work appeared interesting, and he promised to speak about it afterwards with A. N. Kolmogorov. I never knew if this was done, but about a year later at the Functional Analysis Conference (Odessa, 1958), I was greatly surprised to hear in the report of S. V. Fomin about the work of A. N. Kolmogorov in which he introduced $h(T; \xi)$ for a generating partition ξ as the metric invariant for T. (Afterwards, Y. G. Sinai introduced $h(T)$ and proved that $h(T; \xi) = h(T)$ for generating ξ.) S. V. Fomin took my manuscript and sent it to A. N. Kolmogorov. At the beginning of 1959 I again visited A. N. Kolmogorov. At this time we worked in his study for some hours. We sought to understand if my ϵ-entropy was useful. However, it was without any result. As I was leaving, I asked him about my manuscript. He said that after his work, my work had no scientific importance; but he could recommend it for publication in *Uspekhi Mat. Nauk* as a work on the history of the notion of entropy in the study of dynamical systems. I refused such a publication. A. N. Kolmogorov said that his second work on this topic would be published soon, and he would remark on my work in it. This was done [1].

At the time that I became a graduate student, V. P. Potapov had a respite from his fundamental work [2]. He proposed that I continue work on entropy theory. It was a risky choice of topic. There were already two groups of talented mathematicians working intensively in this area, headed by A. N. Kolmogorov and V. A. Rohlin. Naturally, I began by studying the literature on the subject. Only then did I become acquainted with Rohlin's works on measurable partitions. I visited V. P. Potapov's home almost every day the first year, told him about what I had read, and worked in his study. I had dinner there and sometimes supper too. At first it was at the Lastochkina address, where he lived with his first wife (a pianist) and their daughter, and later it was at the

Matrosova address, where he lived with his second wife (a mathematician), her daughter and afterwards their newborn son. One may say that I came to live at V. P.'s (Potapov was called "V. P." by his friends). I felt strongly drawn to him, and I grew fond of him. I believe that the feeling was mutual. V. P. was an interesting man. He fell passionately in love with women. He loved his daughter, music, mathematics, and good wine. He loved talented young people and aided them. In essence, for me he substituted for my father, who died when I was 18 years old. After a time, I began to more critically appreciate the encompassing nature of his influence.

V. P. proposed that I generalize the work of L. M. Abramov on the calculation of the entropy of the automorphisms T of one-dimensional compact commutative groups in the multi-dimensional case. My persistence in finding the solution of this problem was due to the moral support of V. P. and his faith that the problem could be solved. As a result, I did the work in [3] which gives the formula for $h(T)$ in terms of the algebraic invariants of T. The proof of this formula shows that $h(T)$ is not only a metric and an algebraic invariant, but also a topological invariant. This situation raised the problem of topological similarity for such T. After discussing this problem with V. P., he called my attention to Bohr's theorem on the argument of an almost-periodic function. Today I understand that it was not a casual conjecture, because the first dissertation of V. P. was connected with such functions (see [4]). I independently repeated a result of Van Kampen on a generalization of Bohr's theorem, and by this means I obtained results on automorphisms T: 1) two topologically similar automorphisms are algebraically similar, and 2) if these automorphisms are ergodic, then an arbitrary homeomorphism which carries out their topological similarity is the composition of an isomorphism and a shift by a fixed element [5]. Afterwards, this result received an extensive generalization by E. A. Gorin and V. Ya. Lin.

Thus, although my dissertation topic was far from the mathematical interests of V. P., this work did not take place without his participation. I unfortunately did not express gratitude to V. P. for his help in my works [3], [5]. At that time I did not understand that this must be done, and V. P. did not prompt it. Afterwards I understood that my topic was generally not to his taste. At our last meeting he said (not verbatim), "Do you think that anybody could achieve something essential in mathematics on the basis of three axioms?"

Of course, there were useful contacts with Y. G. Sinai (he sent manuscripts to me), participation in Rohlin's seminar (I went on a mission to Leningrad for consultation with V. A. Rohlin), and interaction with V. A. Rohlin, L. M. Abramov, A. M. Vershik, and other participants in Rohlin's seminar. I saw that the specialists were interested in my investigations.

2. A TILT TOWARD OPERATOR THEORY

At the same time that I worked on my dissertation, I attended Kreĭn's lectures on the spectral theory of the string and other differential operators and actively participated in Kreĭn's seminar. At Kreĭn's suggestion, I conducted a teaching seminar on linear prediction of causal stationary sequences and processes. Other participants in this seminar were V. M. Adamjan, H. Langer, V. Javryan, and Sh. Saakyan. My scientific interests began to turn toward operator theory, and they had taken form in this direction by the end of my studies as a graduate student. Thus. M. G. Kreĭn drew me into the area of his scientific interests.

Unfortunately, I had no notion of ℑ-contractive matrix-function theory at that time. I only knew that Potapov's multiplicative theory had connections with characteristic functions in operator theory. I believe V. P. did not pursue the subject at that time. V. P. was at first democratic with respect to my new mathematical interests. It was only later, when V. P. actively began to work on applications of his ℑ-theory to electrical networks and continuation problems, that he began to express his displeasure at my studies on operator theory and to criticize operator theory generally. In the end he succeeded in drawing me into his sphere of interests. But that was later.

At the time that I was producing my first thesis (1964), I was already very involved in my work with V. M. Adamjan on the unitary coupling of semi-unitary operators. It arose after M. G. Kreĭn returned from the Novosibirsk Conference with the preprint of P. Lax and R. Phillips on scattering theory. M. G. Kreĭn raised the problem of establishing a connection between the Lax-Phillips definition of scattering matrix and the definition of that notion by means of wave operators. In Kreĭn's seminar at that time, we began to study the B. Sz.-Nagy and C. Foias works on unitary dilations of contractions. For me the Lax-Phillips and Nagy-Foias works were interesting because they were close to the idea of causal stationary processes and notions of K-flows and K-automorphisms in ergodic theory. I intended to use scattering matrices in ergodic theory, but I did not realize this idea. At first I began to study the Lax-Phillips theory with Kreĭn's graduate student Sh. Saakyan. Soon Kreĭn's graduate student V. M. Adamjan joined us, but Sh. Saakyan went away to his town Erevan and stopped the study. Some time later, our studies in this direction were summed up in [6] and led us to our work with M. G. Kreĭn on Hankel operators and related continuation problems.

3. THE RESULTS OF POTAPOV'S GROUP IN NETWORK THEORY

At the end of the 60's, Potapov's seminars began to operate, at first on $\mathfrak{J}$-theory and related continuation problems and then on electrical network theory too. Participation in these seminars stimulated a change of my interests to the development of dissipative systems theory, $\mathfrak{J}$-inner matrix-function theory, and related continuation problems. We studied the book [7] in the teaching seminar on network theory. For me, networks became a useful model for understanding questions in operator theory, scattering theory, and matrix-function theory. The Darlington method of synthesis of networks with losses was the topic of discussions with V. P. and my participation in his seminars. In 1966, V. P. generalized this method from two ports to 2n-ports. He gave the analytical means for the Darlington method and used the Schur-Potapov algorithm which he developed in the $\mathfrak{J}$-theory. His result was announced at the International Mathematical Congress in 1966 [8].

POTAPOV'S THEOREM. *Any rational real symmetric matrix function* $Z(\lambda)$ $(= \bar{Z}(\bar{\lambda}) = Z^r(\lambda))$ *of order n with* Re $Z(\lambda) > 0$ *for* Re $\lambda \geq 0$ *has a representation in the form*

$$Z(\lambda) = [a_{11}(\lambda)R + a_{12}(\lambda)] [a_{21}(\lambda)R + a_{22}(\lambda)]^{-1}, \tag{1}$$

where $R = I_n$ *and* $A(\lambda) = [a_{ik}(\lambda)]_1^2$ *is a rational real symplectic matrix function of order* $2n$, *that is,*

$$A^r(\lambda)\mathfrak{J}_s A(\lambda) = \mathfrak{J}_s, \qquad \text{where} \qquad \mathfrak{J}_s = \begin{pmatrix} 0 & I_n \\ -I_n & 0 \end{pmatrix},$$

and $A(\lambda)$ *is* $\mathfrak{J}$*-inner for* Re $\lambda > 0$, *that is,*

$$A^*(\lambda)\mathfrak{J}A(\lambda) \leq \mathfrak{J} \qquad \text{for} \quad \text{Re } \lambda > 0, \qquad \text{where} \qquad \mathfrak{J} = \begin{pmatrix} 0 & -I_n \\ -I_n & 0 \end{pmatrix},$$

$$A^*(\lambda)\mathfrak{J}A(\lambda) = \mathfrak{J} \qquad \text{for} \quad \text{Re } \lambda = 0.$$

Potapov's proof gave the representation of $A(\lambda)$ in the form of a product of simple multipliers with the same properties. Some time later, V. P. showed that any $A(\lambda)$ with the stated properties can be represented as such a product, and he established the structure of simple multipliers [9]. Under V. P.'s guidance, A. V. Efimov obtained the realization of an arbitrary multiplier as a transfer matrix function of a $2n \times 2n$ – port of LC type which is lossless and without gyrators; thus the cascade connection of such $2n \times 2n$ – ports gave a realization for an arbitrary rational real symplectic $\mathfrak{J}$-inner matrix function $A(\lambda)$ as the transfer matrix function of a $2n \times 2n$ – port of LC type which is lossless and without gyrators [10].

Let $\mathcal{P}_n$ be the class of positive matrix functions $Z(\lambda)$ of order n, that is, holomorphic $n \times n$ matrix functions satisfying Re $Z(\lambda) \geq 0$ for Re $\lambda > 0$.

Under Potapov's guidance, E. Y. Melamud obtained the representation (1) for an arbitrary rational matrix function of the class $\mathcal{P}_n$. But in (1) she considered $\mathcal{R} = \text{diag}[I_p, 0_{n-p}]$, where $p = \text{rank Re } Z(\lambda)$ for Re $\lambda = 0$, and $A(\lambda) = [a_{ik}(\lambda)]_1^2$ is a rational J-inner matrix function; when $Z(\lambda)$ is real (and symmetric), $A(\lambda)$ can be chosen to be real (and symplectic) too [11]. She obtained this result by the factorization of a matrix function in the form $\varphi^*(-\bar{\lambda})\varphi(\lambda)$, but she did not use the Schur-Potapov algorithm.

Some time earlier, Potapov's graduate student I. V. Kovalishina obtained a representation of an arbitrary rational real J-inner matrix function $A(\lambda)$ as the product of simple multipliers [12]. Another graduate student, T. A. Tovmasyan, obtained a realization of these multipliers as the transfer matrix function of some $2n \times 2n$ – port of LCG type which is lossless and may have gyrators; thus the cascade connection of such $2n \times 2n$ – ports gave him a realization of an arbitrary rational real J-inner matrix function $A(\lambda)$ as a transfer matrix function for a $2n \times 2n$ – port of LCG type which is lossless and may have gyrators [13]. The results of E. Y. Melamud (V. P. Potapov) and T. A. Tovmasyan (A. V. Efimov) give a realization of any rational real (symmetric) matrix function $Z(\lambda)$ of the class $\mathcal{P}_n$, with losses, that is with Re $Z(i\mu) \not\equiv 0$, as the impedance of a $2n$-port of LCGR type (LCR type), obtained by connecting the unit ohm resistance across p outputs and zero ohm resistors (short circuit) across $n - p$ outputs of some lossless $2n \times 2n$ – port of LCG type (LC type); a corresponding $2n \times 2n$ – port is obtained by the cascade connection of simple $2n \times 2n$ – port of the stated type.

A detailed review of the results of Potapov's group on the analysis and synthesis of finite ideal passive networks was given in the paper of A. V. Efimov and V. P. Potapov [14]. Inspired with the successes in this direction, V. P. began to preach in his seminar that J-theory is the "generator" of almost all of the laws of nature. A. V. Efimov was carried away by this idea and attempted to construct a theory of the interaction of elementary particles in such a way, but it was not successful. In essence, V. P. was developing Heisenberg's idea, except that instead of the scattering matrix he focused on the chain scattering (or other data) matrix, which in the physical domain is a contraction too, but in the indefinite metric.

The D-representation (1) together with the multiplicative representation of the corresponding rational J-inner matrix function $A(\lambda)$ was considered by V. P. as a new parametrization of the rational matrix function $Z(\lambda)$ of the class $\mathcal{P}_n$ in the lossy case; it is an alternative to the known additive Nevanlinna formula for $Z(\lambda)\,(\in \mathcal{P}_n)$. In distinction from the latter, this parametrization of $Z(\lambda)$ is defined by a finite number of parameters — the zeros λ_k of the matrix function $Z(\lambda)+Z^*(-\bar{\lambda})$, the poles of $A(\lambda)$, and the matrix parameters of the elementary multipliers for $A(\lambda)$ in (1).

4. DARLINGTON METHOD IN THE GENERAL THEORY OF PASSIVE SYSTEMS

V. P. proposed that I try to generalize the D-representation (1) to the irrational case. I studied this problem simultaneously with the above mentioned investigation of E. Y. Melamud, with mutual influence. An analysis of Potapov's method of obtaining the D-representation (1) in the nondegenerate case revealed that in essence he considered $Z(\lambda)$ as a solution of a Nevanlinna-Pick problem in the class $\mathcal{P}_n$ with the interpolation points λ_k, the zeros of the matrix function $Z(\lambda)+Z^*(-\bar{\lambda})$, and with the values $Z_k = Z(\lambda_k)$. The matrix function $A(\lambda)$ in (1) is a resolvent matrix for this problem: all solutions of the problem may be obtained by the formula (1) in which the matrix R must be replaced by an arbitrary matrix function $R(\lambda)$ of the class $\mathcal{P}_n$ (with values in the relation sense). This resolvent matrix $A(\lambda)$ was obtained by V. P. by the Schur-Potapov algorithm. As soon as I understood this, I started to solve the problem in the scalar case $(n = 1)$ by considering a generalization of the Nevanlinna-Pick problem. Due to the experience of my previous work on scattering matrices and on continuation problems, it was more natural to consider instead of $\mathcal{P}_n$ the class $B_n = B_{n \times n}$, where $B_{n \times m}$ is the class of holomorphic contraction $n \times m$ matrix functions $S(\lambda)$ defined for Re $\lambda > 0$, and $Z = (I - S)(I + S)^{-1}$.

For $S \in B_{n \times m}$, the D-representation corresponding to (1) is written in the form

$$S(\lambda) = [W_{11}(\lambda)\mathcal{E} + W_{12}(\lambda)][W_{21}(\lambda)\mathcal{E} + W_{22}(\lambda)]^{-1}. \qquad (2)$$

where $\mathcal{E} = \mathrm{const.}\,(\in B_{n \times m})$ if $m \neq n$, $\mathcal{E} = \mathrm{diag}\,[0_p, I_{n-p}]$ if $m = n$,

$$p = \mathrm{rank}\,[I - S^*(-\bar{\lambda})S(\lambda)],$$

and $W(\lambda) = [W_{ik}(\lambda)]_1^2$ is a j-inner matrix function defined for Re $\lambda > 0$, $j = \mathrm{diag}\,[I_n, -I_m]$.

In the nondegenerate case $(p = m)$, the D-representation (2) is equivalent to the D-representation of $S(\lambda)$ in the form of the block $S_{12}(\lambda)$ of an inner matrix function

$\tilde{S}(\lambda) = [S_{ik}(\lambda)]_1^2$ of order $n + p$, that is,

$$\tilde{S}(\lambda) = \begin{pmatrix} S_{11}(\lambda) & S_{12}(\lambda) \\ S_{21}(\lambda) & S_{22}(\lambda) \end{pmatrix}, \qquad S_{12}(\lambda) = S(\lambda). \tag{3}$$

The matrices W and $\tilde{S}$ in formulas (2) and (3) are connected by the Potapov-Ginsburg transformation:

$$\tilde{S} = [P_- + P_+ W][P_+ + P_- W]^{-1}, \qquad P_{\pm} = \frac{1}{2}[I \pm j].$$

Comparatively quickly, I understood that for D-representations to exist for the given $S(\lambda)$ (or $Z(\lambda)$), it is necessary that quasi-continuations exist in the left half-plane, that is,

$$S(i\mu + 0) = S(i\mu - 0) \qquad \text{a.e.,}$$

where $S(i\mu - 0)$ are the boundary values for $S(\lambda)$, which is a meromorphic matrix function of bounded Nevanlinna characteristic in the left half-plane. In honor of Potapov, I denoted the class of such matrix functions (or operator functions) S as $B\Pi$ (the corresponding class of matrix functions $Z(\lambda)$ was defined as $\mathcal{P}\Pi$). I turned my attention to the class $B\Pi$ with knowledge of the Douglas-Shapiro-Shields work on noncyclic vectors for the *-shift. It was more difficult to establish that the condition $S \in B\Pi$ ($Z \in \mathcal{P}\Pi$) is also sufficient for the existence of D-representations. For this I first considered S_0 from $B\Pi$ with $\|S_0\|_\infty < 1$, and in the scalar case I obtained the representation (2) by considering the generalized Nevanlinna-Pick problem: find all S for which

$$b_0^{-1}(\lambda)[S(\lambda) - S_0(\lambda)] \in H^\infty. \qquad S \in B,$$

where b_0 is an inner function. All of the solutions of this problem may be described by formula (2) with some j-inner matrix function $W(\lambda) = [W_{ik}(\lambda)]_1^2$, the resolvent matrix of the problem, and an arbitrary function $\mathcal{E} \in B$. If in that problem b_0 is such that

$$b_0(\lambda)S_0^*(-\bar{\lambda})\left[I - S_0(\lambda)S_0^*(-\bar{\lambda})\right]^{-1} \in H^\infty,$$

(or for the corresponding $Z(\lambda)$ from $\mathcal{P}\Pi$,

$$b_0(\lambda)\left[Z(\lambda) + Z^*(-\bar{\lambda})\right]^{-1} \in H^\infty,)$$

then the given $S_0(\lambda)$ is defined by considering formula (2) with $\mathcal{E} = \text{const.}$, and we may take $\mathcal{E} = 0$. If $S_0 \in B\Pi$ but $\|S_0\|_\infty = 1$, we may consider the D-representation (2) for ρS_0 with $0 < \rho < 1$ and after this pass to the limit by letting $\rho \uparrow 1$.

For matrix functions and operator functions of class $B\Pi$ I obtained the D-representation (3) by the Rosenblum-Rovnyak factorization theorem on the representation of operator functions in the form $\varphi^*(-\bar\lambda)\varphi(\lambda)$ with outer $\varphi(\lambda)$. The factorization method allowed me to describe all the D-representations (3) for the given S ($\in B\Pi$) and furthermore gave D-representations (3) which are minimal and optimal in a natural sense. When I was deriving the D-representations (1) and (2) by the factorization method in the degenerate case, I used the manuscript of E. Y. Melamud which was published later [11].

As soon as I generalized the D-representation (1) for the matrix function $Z(\lambda)$ of the class $\mathcal{P}_n$, Potapov raised the problem of applying this result to the theory of the string. V. P. thought that it was unnatural to consider singular strings (with infinite length or mass) and that the impedance $Z(\lambda)$ of such a string is also the impedance of a regular string with friction on the other end of the string. He thought that such a result might be obtained by the Darlington method. After more profound study of Kreĭn's work on the theory of the string, I realized that there is a much more general problem on the impedances $Z(\lambda)$ of $2n$-ports which are n ideal regular lossless transmission lines with distributed parameters and with losses on the other ends of the lines. This means that $\widehat{u}(\lambda) = Z(\lambda)\widehat{i}(\lambda)$, where $\widehat{u}(\lambda)$ and $\widehat{i}(\lambda)$ are the Laplace transforms of $u(t) = \mathrm{col}[u_k(t)]_1^n$ and $i(t) = \mathrm{col}\,[i_k(t)]_1^n$, $u(t) = u(0,t)$, $i(t) = i(0,t)$, and $u(x,t)$ and $i(x,t)$ are the solutions of the regular canonical differential system with dissipative boundary conditions:

$$\frac{\partial}{\partial x}\begin{pmatrix} u(x,t) \\ i(x,t) \end{pmatrix} = -\Im H(x)\frac{\partial}{\partial t}\begin{pmatrix} u(x,t) \\ i(x,t) \end{pmatrix}, \quad (0 \le x \le l), \quad u(l,t) = Ri(l,t) \qquad (4)$$

$$u(x,0) = 0, \qquad i(x,0) = 0,$$

where $R > 0$, $H(x) \ge 0$, $\int_0^l \|H(x)\|\, dx < \infty$, and R is a matrix of order n and $H(x)$ is a matrix function of order $2n$. I obtained the following results.

THEOREM. *A matrix function $Z(\lambda)$ which is meromorphic in the complex plane and of class $\mathcal{P}_n$ is the impedance for some problem (4) if and only if $[Z(\lambda) + Z^*(-\bar\lambda)]^{-1}$ is an entire matrix function of exponential class and Cartwright class and $Z(\lambda)$ has no poles on the imaginary axis.*

THEOREM. *A matrix function $Z(\lambda)$ is the impedance for some problem (4) with real R ($= \bar R > 0$) and real block diagonal $H(x)$ ($=$ diagonal $[C(x), L(x)] = \overline{H(x)} \ge 0$) if and only if in addition to the conditions of the last theorem, $Z(\lambda)$ is a real and symmetric matrix function.*

In the scalar case ($n = 1$), this theorem gives criteria for the impedance of a regular string with friction on the other end of the string.

Essential ingredients in the proof of the first theorem are the Potapov theorem on the multiplicative representation of an entire $\mathfrak{J}$-inner matrix function $A(\lambda)$ and the Kreĭn theorem on the representation of an entire matrix function in the form $\varphi^*(-\bar{\lambda})\varphi(\lambda)$ with an entire matrix function $\varphi(\lambda)$. For the second theorem, I used the Kreĭn extension of the Potapov theorem on multiplicative representation to the case of real and symplectic entire $\mathfrak{J}$-inner matrix functions $A(\lambda)$.

The D-representation (3) prompted me to note the connection between the Darlington method and the theory of characteristic functions of operators and the Lax-Phillips scattering theory. At that time, A. A. Nudel'man reported in Kreĭn's seminar on R. Kalman's work on the minimal dissipative realization for rational functions of the class $\mathcal{P}_1$. On the other hand, I repeatedly heard criticism of the theory of characteristic functions from V. P. in his seminar. A fundamental reason for his view was this: it is not natural that a function $S(\lambda)$ of the class B_1, in the rational non-inner case, is the characteristic function of a dissipative operator acting in an infinite-dimensional space. He compared this with the situation in network theory where rational functions $Z(\lambda)$ of the class $\mathcal{P}_1$ are impedances of finite networks. Soon I understood that V. P. identified the notions of dissipative operator and dissipative system. Just as in operator theory the function $S(\lambda)$ of class B is considered as the transfer function of a conservative scattering system with dissipative basic operator, in network theory the rational functions $S(\lambda)$ ($\in B$) and $Z(\lambda)$ ($\in \mathcal{P}$) in the lossless case are considered as the transfer functions of dissipative systems. As soon as I understood that, I concentrated my attention on the investigation of dissipative systems. Thus V. P.'s criticism of operator theory, which was often tendentious and subjective, had a positive influence on me.

My basic results on the Darlington method and related problems (the D-representations (1), (2), (3), the applications to canonical systems, and the connections with unitary dilation theory and the Kalman theory of minimal dissipative realizations) were announced in the article [15] (1971). A detailed account of the analytical results on the D-representations (1) and (2) was given in the works [16], [17]; on their applications to canonical differential systems with dissipative boundary conditions and to the theory of the string, in the work [17]; on the D-representation (3) and its connection with conservative and dissipative scattering theory, in the works [18] and [19]; and on connections with rational approximation theory, in the work [20]. The general results on passive realizations, in particular on minimal passive realizations, was given in the papers [21], [22]. In the works [23] and [19], I considered minimal and optimal passive realizations of the transfer functions $Z(\lambda)$ ($\in \mathcal{P}$) and $S(\lambda)$ ($\in B$), that is, passive realizations with minimal state space and minimal norm ("energy") of the inner states.

V. P. and his group and I developed the Darlington method independently of the works [24], [25] of V. Belevich, in which he generalized the Darlington result on multipoles with losses by considering the rational D-representation (3). He obtained all such representations by the factorization method. Proceeding from this work, P. Dewilde [26] obtained criteria for the existence of the D-representation (3) in the scalar case, and R. Douglas and J. W. Helton [27] generalized the result to operator functions: they showed independently of me that the condition $S(\lambda) \in B\Pi$ is sufficient for the existence of the D-representation (3).

In the work [28] I gave a review of results on the theory of passive linear systems and related problems.

5. REGULAR j-INNER MATRIX FUNCTIONS AND RELATED GENERALIZED BITANGENTIAL PROBLEMS

As soon as it was clear that in the D-representation (1) in the nondegenerate case $A(\lambda) = [a_{ik}(\lambda)]_1^2$ is the resolvent matrix of a Nevanlinna-Pick problem, V. P. began to study the connection between Blaschke-Potapov products and the Nevanlinna-Pick problem. He successfully solved this problem in [29]. I. P. Fedčina generalized his result on the tangential Nevanlinna-Pick problem in [30]. But this problem was connected with Blaschke-Potapov products for which the elementary multipliers have poles in one half-plane, right or left. When these poles are in both half-planes, such Blaschke-Potapov products are connected with the bitangential Nevanlinna-Pick problem in which data are given for tangential interpolation of $Z(\lambda)$ and $Z^*(\lambda)$. In the class $B_{n \times m}$ such bitangential problems may be formulated in this way: Let $S_0 \in B_{n \times m}$, $b_1 \in B_n$, $b_2 \in B_m$ be given, where b_1 and b_2 are Blaschke-Potapov products. Find the matrix functions S such that

$$b_1^{-1}(S - S_0)b_2^{-1} \in H_{n \times m}^\infty, \qquad S \in B_{n \times m}. \tag{5}$$

In our paper with L. A. Simakova [31], we proved that in the j-metric case every convergent Blaschke-Potapov product $W(\lambda)$ is a j-inner matrix function.

For an arbitrary j-inner matrix function $W(\lambda) = [W_{ik}(\lambda)]_1^2$, we have

$$(S_{11}(\lambda) =) \; [W_{11}^*(-\bar\lambda)]^{-1} \in B_n, \qquad (S_{22}(\lambda) =) \; W_{22}^{-1}(\lambda) \in B_m,$$

and therefore

$$W_{11}(\lambda) = b_1(\lambda)p_-(\lambda), \qquad W_{22}(\lambda) = b_2^{-1}(\lambda)p_+(\lambda), \tag{6}$$

where b_1 ($\in B_n$) and b_2 ($\in B_m$) are inner matrix functions and $p_-^*(-\bar\lambda)$ and $p_+(\lambda)$ are outer matrix functions, $[p_-^*(-\bar\lambda)]^{-1} \in B\Pi$, $p_+^{-1}(\lambda) \in B\Pi$.

The j-inner matrix function $W(\lambda)$ is called singular if it is an outer matrix function in the class of matrix functions with bounded Nevanlinna characteristic, that is, if and only if b_1 and b_2 in (6) are constant matrices. The j-inner matrix function $W(\lambda)$ is called regular if it has no nonconstant singular right divisor in the class of j-inner matrix functions. The Blaschke-Potapov product is a singular j-inner matrix function if and only if all of its elementary multipliers have poles on the imaginary axis, and it is a regular j-inner matrix function if and only if none of its poles are on the imaginary axis.

The investigations of the Darlington method and j-inner matrix functions led me to the generalized bitangential Schur-Nevanlinna-Pick problem (5) with arbitrary given inner matrix function b_1 ($\in B_n$) and b_2 ($\in B_m$) [32], [33].

Let $B_{S_0;b_1,b_2}$ be the set of all solutions of the problem (5). Problem (5) is completely indeterminate if and only if

$$\forall \xi \ (\in C^m, \ \xi \neq 0) \quad \exists \, S \ (\in B_{S_0;b_1,b_2}) \ : \ S(\lambda_0)\xi \neq S_0(\lambda_0)\xi,$$

where λ_0 is a fixed point with Re $\lambda_0 > 0$ and

$$\det b_1(\lambda_0) \det b_1(\lambda_0) \neq 0.$$

The connection between completely indeterminate problems (5) and j-inner matrix functions was established in the works [32, 33].

THEOREM. *Let* $W = [W_{ik}]_1^2$ *be an arbitrary j-inner matrix function, $S_0 = W_{12}W_{22}^{-1}$, and let b_1 and b_2 be the inner matrix functions in (6). Then the problem (5) taken with these S_0, b_1, and b_2 is completely indeterminate, and the formula (2) with arbitrary $\mathcal{E} = \mathcal{E}(\lambda) \in B_{n \times m}$ gives the solution S to this problem. Formula (2) gives all solutions of the problem (5), that is, W is the resolvent matrix for the problem (5), if and only if W is regular. For an arbitrary completely indeterminate problem (5) there exists a regular resolvent matrix W with b_1 and b_2 in (6) the same as in problem (5). This matrix function W is determined by the problem (5) up to a constant j-unitary right factor.*

THEOREM. *Let* $S_0 \in B_{n \times m}\Pi$ *and* $\|S_0\| < 1$. *Then the solution S_0 to the problem (5) is obtained by the formula (2) with $\mathcal{E} = $ constant if and only if*

$$b_2(\lambda)S_0^*(-\bar{\lambda})\left[I - S_0(\lambda)S_0^*(-\bar{\lambda})\right]^{-1} b_1(\lambda) \in H_{n \times m}^\infty.$$

If $S \in B_{n \times m}\Pi$ and $\|S\| < 1$, all D-representations (2) for S may be obtained in this way.

My graduate student L. A. Simakova showed [34, 35] that if $W = [W_{ik}]_1^2$ is a meromorphic matrix function defined for Re $\lambda > 0$ with $\det W(\lambda) \not\equiv 0$ and blocks W_{11}

and W_{22} of order n and m such that formula (2) transforms an arbitrary $\mathcal{E}$ in $B_{n \times m}$ into an element S of $B_{n \times m}$, then there exists a meromorphic scalar function $\rho(\lambda)$ for Re $\lambda > 0$ such that $\rho(\lambda)W(\lambda)$ is a j-contractive matrix function for Re $\lambda > 0$. If in the formula (2) all inner matrix functions $\mathcal{E}$ are transformed into inner matrix functions S, then $\rho(\lambda)W(\lambda)$ is a j-inner matrix function.

In work with L. Z. Grossman [36, 37], we developed the Kreĭn theory of the resolvent matrix and Livšic' theory of characteristic functions. It follows from this investigation that an arbitrary j-contractive matrix function $W(\lambda)$ defined for Re $\lambda > 0$ is the resolvent matrix for some abstract interpolation problem as considered by V. E. Katsnelson and his graduate students P. M. Yuditskiĭ and A. Ya. Kheifets [38, 39].

V. P. advocated the point of view that for completely indeterminate continuation problems in the class $B_{n \times m}$ (or $\mathcal{P}_n$) with resolvent matrix $W(\lambda)$ $(A(\lambda))$, information about various properties of the set of all solutions must be obtained directly by considering the resolvent matrix.

In this way I obtained general results about solutions with maximal jump of spectral functions for $Z(\lambda)$ $(\in \mathcal{P}_n)$ [33, 40], and in works with M. G. Kreĭn [41, 42] we obtained general results on solutions with maximal entropy.

In the articles [43, 44, 45] I announced some results on the generalized bitangential problem of Carathéodory-Nevanlinna-Pick-Kreĭn in the class $\mathcal{P}_n$ corresponding to the problem (5). Here are given results on Kreĭn continuation problems for helical matrix functions and positive definite matrix functions and on more general continuation problems for such matrix functions.

I believe that the influence of Potapov and Kreĭn will continue to show in my future work, and more than a little.

Finally, I wish to thank James Rovnyak and Harry Dym for helping to turn my translation into more recognizable English.

REFERENCES

[1] A. N. Kolmogorov, *On the entropy of unit time as the metric invariant automorphisms*, Dokl. Akad. Nauk SSSR 124, no. 4 (1959), 754–755.

[2] V. P. Potapov, *The multiplicative structure of J-contractive matrix functions*, Trudy Moskov. Mat. Obsc. 4 (1955), 125–236; Amer. Math. Soc. Transl. (2) 15 (1960), 131–243. MR 17, 958.

[3] D. Z. Arov, *Calculation of entropy for a class of group endomorphisms*, Zap. Meh.-Mat. Fak. Harkov. Gos. Univ. i Harkov. Mat. Obsc. 30 (4) (1964), 48–69. MR 35#4368.

[4] V. P. Potapov, *On the divisors of almost-periodic polynomials*, Slovnik Trudov Inst. Mat. Akad. Nauk Ukr. SSR, no. 12, 1949, 36–81.

[5] D. Z. Arov, *Topological similitude of automorphisms and translations of compact commutative groups*, Uspekhi Mat. Nauk 18, no. 5, (1963), 133-138. MR 27#5858.

[6] V. M. Adamjan and D. Z. Arov, *Unitary couplings of semi-unitary operators*, Akad. Nauk Armjan. SSR Dokl. 43, no. 5, (1966), 257–263. MR 34#8201.

[7] S. Seshu and M. B. Rid, *Linear Graphs and Electrical Networks*.

[8] V. P. Potapov, *Multiplicative representations of analytic matrix-valued functions*, Abstracts of Brief Scientific Communications, Internat. Congr. of Math., Moscow, 1966, Section 4, 74–75.

[9] V. P. Potapov, *General theorems of structure and the splitting of elementary multipliers of analytic matrix-functions*, Akad. Nauk Armjan. SSR Dokl. 48, no. 5, (1969), 257–263. MR 57#16627.

[10] A. V. Efimov, *Realization of reactive j-contractive matrix functions*, Izv. Armjan. SSR 5, no. 1 (1970), 54–63.

[11] E. Y. Melamud, *On a generalization of Darlington's theorem*, Izv. Akad. Nauk Armjan. SSR, Ser. Mat., 7 (1972), 183–195.

[12] I. V. Kovalishina, *The multiplicative structure of analytical reactive matrix functions*, Izv. Armjan. SSR Mat. 1, no. 2, (1966), 138-146.

[13] T. A. Tovmasyan, *On the elementary and primal multipliers of j-contractive real matrix functions*, Uchen. Zap. Erevan Univ. 1 (1971), 11–26.

[14] A. V. Efimov and V. P. Potapov, *J-expansive matrix-valued functions, and their role in the analytic theory of electrical circuits*, Uspekhi Mat. Nauk 28, no. 1 (169), (1973), 65–130; Russian Math. Surveys 28, 1973, no. 1, 69–140. MR 52#15090.

[15] D. Z. Arov, *Darlington's method in the study of dissipative systems*, Dokl. Akad. Nauk SSSR 201 (1971), 559-562; Soviet Physics Dokl. 16 (1971/72), 954–956. MR 55#1127.

[16] D. Z. Arov, *Darlington realization of matrix-valued functions*, Izv. Akad. Nauk SSSR Ser. Mat. 37 (1973), 1299–1331; Math. USSR-Izv. 7 (1973), 1295–1326. MR 50#10287.

[17] D. Z. Arov, *Realization of a canonical system with a dissipative boundary condition at one end of the segment in terms of the coefficient of dynamical compliance*, Sibirsk. Mat. Z. 16, no. 3, (1975), 440–463. MR 57#12872.

[18] D. Z. Arov, *Unitary couplings with losses (a theory of scattering with losses)*, Funktsional. Anal. i Prilozhen. 8, no. 4, (1974), 5–22. MR 50#10864.

[19] D. Z. Arov, *Stable dissipative linear stationary dynamical scattering systems*, J. Operator Theory 2, no. 1, (1979), 95–126. MR 81g:47007.

[20] D. Z. Arov, *An approximation characteristic of functions of the class BΠ*, Funktsional. Anal. i Prilozhen. 12, no. 2, (1978), 70–71. MR 58#11420.

[21] D. Z. Arov, *Scattering theory with dissipation of energy*, Dokl. Akad. Nauk SSSR 216, no. 4, (1974), 713–716. MR 50#14287.

[22] D. Z. Arov, *Passive linear steady-state dynamical systems*, Sibirsk. Mat. Zh. 20, no. 2, (1979), 211–228. MR 80g:93031.

[23] D. Z. Arov, *Optimal and stable passive systems*, Dokl. Akad. Nauk SSSR 247, no. 2, (1979), 265–268. MR 80k:93036.

[24] Y. Belevich, *Synthèse des rèseaux electriques passifs à n paires de bornes de matrice de rèpartition prèdèterminèe*, Ann. Telecomm 6 (11) (1951), 302–312.

[25] V. Belevich, *Classical Network Theory*, San Francisco, 1978.

[26] P. Dewilde, *Roomy scattering matrix synthesis*, Technical Report, Berkeley, 1971.

[27] R. G. Douglas and J. W. Helton, *Inner dilations of analytic matrix functions and Darlington synthesis*, Acta Sci. Math. (Szeged) 34 (1973), 61–67. MR 48#900.

[28] D. Z. Arov, *Some problems in the theory of linear stationary passive systems*, Operators in indefinite metric spaces, scattering theory and other topics (Bucharest, 1985), Oper. Theory: Adv. Appl. OT24, Birkhäuser, Basel, 1987, pp. 17-27. MR 88j:93003.

[29] I. V. Kovalishina and V. P. Potapov, *An indefinite metric in the Nevanlinna-Pick problem*, Akad. Nauk Armjan. SSR Dokl. 59, no. 1, (1974), 17–22. MR 53#13577.

[30] I. P. Fedčina, *The tangential Nevanlinna-Pick problem with multiple points*, Akad. Nauk Armjan. SSR Dokl. 61 (1975), 214–218. MR 53#13576.

[31] D. Z. Arov and L. A. Simakova, *The boundary values of a convergent sequence of J-contractive matrix-valued functions*, Mat. Zametki 19, no. 4, (1976), 491-500. MR 58#6278.

[32] D. Z. Arov, *γ-generating matrices, j-inner matrix functions and related extrapolation problems. I* Teor. Funktsii Funktsional. Anal. i Prilozhen. 51 (1989), 61–67; Part II, ibid. 52 (1989), 103–109; Part III, ibid. 53 (1990), 57–64; J. Soviet Math. 52 (1990), 3487-3491; 52 (1990), 3421-3425; 58 (1992), 532-537.

[33] D. Z. Arov, *Regular j-inner matrix functions and related continuation problems*, Oper. Theory: Adv. Appl. OT43, Birkhäuser, Basel, 1990, pp. 63–87.

[34] L. A. Simakova, *On plus-matrix-functions and related continuation problems*, Mat. Issled., Kishinev, 9, no. 2, (1974), 149–171.

[35] L. A. Simakova, *On meromorphic plus-matrix-functions*, Mat. Issled., Kishinev, 10, no. 1, (1975), 287–292.

[36] D. Z. Arov and L. Z. Grossman, *Scattering matrices in the theory of extensions of isometric operators*, Dokl. Akad. Nauk SSSR 270, no. 1, (1983), 17–20. MR 85c:47008.

[37] D. Z. Arov and L. Z. Grossman, *Scattering matrices in the theory of unitary extensions of isometric operators*, Math. Nachr. 157 (1992), 105–123.

[38] V. E. Katsnelson, A. Ya. Kheifets, and P. M. Yuditskiĭ, *An abstract interpolation problem and the theory of extensions of isometric operators*, Operators in Function Spaces and Problems in Function Theory, Collected Scientific papers, Kiev, Naukova Dumka, 1987, 83–96.

[39] A. Ya. Kheifets, *Parseval's equality in the abstract interpolation problem and the coupling of open systems. I*, Teor. Funktsii Funktsional. Anal. i Prilozhen. 49 (1988), 112–120; Part II, ibid. 50 (1988), 98–103; J. Soviet Math. 49, no. 4, 1114–1120 (Part I) and 49, no. 6, 1307–1310 (Part II).

[40] D. Z. Arov, *The Carathéodory theorem for matrix-functions, qthe maximal jump of spectral functions in continuation problems*, Mat. Zametki 48, no.3, (1990), 3-11.

[41] D. Z. Arov and M. G. Kreĭn , *The problem of finding the minimum entropy in indeterminate problems of continuation*, Funktsional. Anal. i Prilozhen. 15, no. 2, (1981), 61–64. MR 84m:47027.

[42] D. Z. Arov and M. G. Kreĭn , *Calculation of entropy functionals and their minima in indeterminate continuation problems*, Acta Sci. Math. (Szeged) 45 (1983), 33–50. MR 84h:30050.

[43] D. Z. Arov, *The generalized bitangential Carathéodory-Nevanlinna-Pick problem and $(j,\mathfrak{J})$-inner matrix functions*, Internat. Workshop on Algorithms and Parallel VLSI Architectures, June 10–16, 1990, Abbaye des Premontres, Pont-a-Mousson-France, Part B, 179–183.

[44] D. Z. Arov, *$(j,\mathfrak{J})$-inner matrix functions and the generalized bitangential Carathéodory-Nevanlinna-Pick-Kreĭn problem*, WOTCA Workshop on Operators and Complex Analysis, June 11–14, 1991, 4–5, Abstracts, Hokkaido University, Sapporo, Japan.

[45] D. Z. Arov, *The generalized bitangential Carathéodory-Nevanlinna-Pick problem and $(j,\mathfrak{J})$-inner matrix functions*, Izvestiya Rossiman Akad. Nauk, no. 1 (1993), to appear.

Odessa Pedagogical Institute
26 Komsomol Str.
270020 Odessa
Ukraine

Operator Theory:
Advances and Applications, Vol. 72
© 1994 Birkhäuser Verlag Basel

THE DEVELOPMENT OF SOME OF V.P. POTAPOV'S IDEAS
THE GEOMETRIC THEORY OF OPERATORS IN SPACES
WITH INDEFINITE METRIC

T.Ja. Azizov and E.I. Iohvidov

The results of V.P. Potapov, relating to geometric questions of matrix theory in spaces with an indefinite metric, are concentrated mainly in his fundamental research [16]. His ideas were the basis of investigations of a series of authors; in some cases without these authors being aware of the fact that their research was rooted in Potapov's work (as, for example. Colpa [3] and Ikramov [6]). Our survey does not claim to be a complete reflection of V.P. Potapov's ideas in the field mentioned and it is concentrated around three objects closest to us: *the criterion of doubleness of J-nonexpansive operators, J-polar representation of plus-operators and factorization of J-nonnegative operators.*

1. In the initial paper of Potapov [16] a remarkable property of *J-nonexpansive matrices* was discovered. Recall that a matrix A is said to be J-nonexpansive if

$$A^* J A \leq J, \tag{1}$$

where

$$J = \begin{pmatrix} I_p & 0 \\ 0 & -I_q \end{pmatrix}, \qquad p + q = \dim L < \infty,$$

and where L is a complex Euclidean space.

It turned out that the matrix A^*, conjugate to the J-nonexpansive matrix A, is also J-nonexpansive, which is equivalent to the matrix inequality

$$A J A^* \leq J \tag{2}$$

In other words, it turned out that the conditions (1) and (2) are equivalent ([16], Chapter 2, Theorem 7). Later in [17] another proof of this statement was given.

In a lot of papers this result of V.P. Potapov (just as many others) was extended to the case of operators in an infinite dimensional space with an indefinite metric, and the authors of these papers almost always assumed that the operator A is bounded and that it is defined on the whole space. In the papers of Ginzburg [4] and [5] a concept of a doubly-J-nonexpansive operator in a Krein space was introduced (using up-to-date terminology), and the most important properties of these operators were established. Recall that a krein space is a Hilbert space $\mathcal{H}$ with the additional structure:

$$\mathcal{H} = \mathcal{H}_+ \oplus \mathcal{H}_-, \quad \mathcal{H}_\pm = P_\pm \mathcal{H}, \quad P_+ + P_- = I, \quad P_\pm = P_\pm^2 = P_\pm^*,$$

and with a J-metric

$$[x, y] = (Jx, y), \qquad J = P_+ - P_-,$$

An operator $A \in B(\mathcal{H})$ was called *doubly-J-nonexpansive*, if it satisfies two conditions:

$$[Ax, Ax] \leq [x, x] \quad \forall x \in \mathcal{H}, \tag{3}$$

$$[A^c x, A^c x] \leq [x, x] \quad \forall x \in \mathcal{H}, \tag{4}$$

where A^c is the J-conjugate of A ($A^c = JA^*J$). In addition, in the papers [4] and [5] it was shown, that the conditions (3) and (4), generally speaking, are not equivalent, and besides the following criterion of doubleness was proved:

In order that a J-nonexpansive operator $A \in B(\mathcal{H})$ is doubly-J-nonexpansive, it is necessary and sufficient that at least one of the following equivalent conditions is realized

1) *The operator A^* is J-nonexpansive;*

2) $\lambda = 0$ *is a regular point of the operator* $P_- A P_- : 0 \in \rho(P_- A P_-)$;

3) $0 \in \rho(P_+ - P_- A)$.

In the same place, as a corollary of the above mentined criterion, it was noted that *in the Pontrjagin space Π_κ, where $\kappa = \dim \mathcal{H}_- < \infty$, every J-nonexpansive operator is doubly-J-nonexpansive.*

In the papers of I. Iohvidov [10], [11] and [12] the criterion of Ju. Ginzburg in its sufficient part was extended to the case of an unbounded J-nonexpansive operator A such that $\mathcal{D}(A) = \mathcal{H}$. It should be noted also, that in the above mentioned papers [4], [5], [10]-[12] the criteria of the doubleness were given not only in terms of the J-nonexpansive operator A itself, but in terms of fractional-linear transformation

$$B = (P_- + P_+ A)(P_+ + P_- A)^{-1} \tag{5}$$

of this operator too. It was shown that such a transformation has meaning for every J-nonexpansive operator and the operator B is a contraction. Previously the transformation (5) along with fractional-linear transformations of more general form had been widely used by V.P. Potapov for studying operators in finite dimensional spaces with a J-metric [16].

New approaches to the doubleness problem were suggested by M.G. Krein and Ju.L. Shmul'jan in [13] and [14]. Let us remind that the symbol $\mathcal{M}_-$ denotes the set of all maximal nonpositive subspaces of a Krein space (see [2]). In particular, in [13] it was proved:

In order that a J-nonexpansive operator $A \in B(\mathcal{H})$ is a doubly-J-nonexpansive operator, it is necessary that an arbitrary subspace $\mathcal{L} \in \mathcal{M}_-$ is mapped by the operator A on a subspace from the set $\mathcal{M}_-$. Furthermore, it is sufficient that at least one subspace $\mathcal{L} \in \mathcal{M}_-$ is mapped by A on a subspace from the set $\mathcal{M}_-$.

Another criterion, proposed in [14], runs as follows:

A J-nonexpansive operator $A \in B(\mathcal{H})$ is doubly-J-nonexpansive if and only if the following two conditions hold:

1) The spectrum of the operator $A^c A$ is nonpositive;

2) The space Ker A^c is uniformly positive [2].

Note that formerly in the papers [4] and [5] Ju.P. Ginzburg established that *every doubly-J-nonexpansive operator $A \in B(\mathcal{H})$ possesses properties 1) and 2).*

In E. Iohvidov's paper [7] a criterion of doubleness was obtained for a J-nonexpansive operator $\widetilde{V}$, which is an extension of a J-isometric operator V

$$([Vx, Vy] = [x, y] \qquad \forall x, y \in \mathcal{D}(V)),$$

satisfying the next conditions:

   1) The operator V is invertible, bounded and closed;
   2) $\dim \mathcal{D}(V)^\perp = \dim \mathcal{R}(V)^\perp < \infty$;
   3) The inertia indices [2] of the subspaces $\mathcal{D}(V)^\perp$ and $\mathcal{R}(V)^\perp$ coincide.

The inertia indices of a finite dimensional subspace are given by the number of positive, negative, and zero squares of the indefinite inner product on this subspace. It turned out, that in this case the operator $\widetilde{V}$ is *doubly-J-nonexpansive if and only if*, at least one of the following four equivalent conditions holds:

   1) $\mathcal{D}(\widetilde{V}) = \mathcal{H}$;
   2) $(P_+ + P_-\widetilde{V})\mathcal{D}(\widetilde{V}) = \mathcal{H}$;
   3) $\mathcal{D}(\widetilde{U}) = \mathcal{H}$;
   4) $(P_+ + P_-\widetilde{U})\mathcal{D}(\widetilde{U}) = \mathcal{H}$.

The operator $\widetilde{U}$, appearing in this criterion, is a transformation (5) of the operator $\widetilde{V}$, i.e.

$$\widetilde{U} = (P_- + P_+\widetilde{V})(P_+ + P_-\widetilde{V})^{-1}.$$

In the paper of E. Iohvidov [8] a criterion of generalized doubleness for J-nonexpansive operator, having dense domain in Krein space and allowing closure, was established. In the paper [9] of the same author more general classes of operators then J-nonexpansive (classes Γ_α, where $\alpha > 1$) were considered, and, besides, it managed to get rid of the "superfluous" conditions of the closedness of an operator. An operator V belongs, by the definition, to the class Γ_α ($\alpha > 0$), if the following inequality is valid:

$$|P_+Vx|^2 - \alpha|P_-Vx|^2 \le \alpha|P_+x|^2 - |P_-x|^2 \qquad \forall x \in \mathcal{D}(V).$$

We note that the class Γ_1 coincides with the class of all J-nonexpansive operators, and, in addition, $\Gamma_\alpha \subseteq \Gamma_\beta$, if $\alpha \le \beta$. The criterion, proved in [9], runs as follows:

   Let $V \in \Gamma_\alpha$ $(\alpha > 0)$, $\mathcal{D}(V) = \mathcal{H}$. *Then the next three conditions are equivalent:*

   1) $V^c \in \Gamma_\alpha$.

2) $\mathrm{Ker}\,(P_+ + P_- V^c) = \{0\}$.

3) $(P_+ + P_- V)\mathcal{D}(V) = \mathcal{H}$.

Some generalizations of the concept of a doubly-J-nonexpansive operator were proposed in V.S. Ritzner's paper [18]. We note in conclusion the peculiar approach to the doubleness problem, developed by S.A. Kuzhel [15]. The essence of his approach consists of finding and describing all subspaces of the Krein space, where an operator, conjugate to a J-nonexpansive, is also a J-nonexpansive operator.

Summing up the considerations of the present section, we may affirm with absolute reason that the initial theorem of the V.P.Potapov ([16], Chapter 2, Theorem 7) was, on the one hand, a powerful stimulus for many investigations and generalizations, and, on the other hand, promoted the development of new important and useful concepts in the theory of operators in spaces with indefinite metric (in particular, for the classes of doubly-J-nonexpansive operators, doubly-J-noncontractive operators, doubly strict plus-operators and others).

2. At the same time the above mentioned Theorem 7 in [16] was broadly utilised. In particular, this theorem was used for a construction of the J-polar represen-tation and the J-modulus of a J-nonexpansive matrix, and besides for en elucidation of different connections between the matrix $J - T^* J T$ (J-form) and the J-modulus of J-nonexpansive matrix T. Just with the help of these results a theorem about the product of J-moduli was obtained, which is very important for a proof of convergence of a product of elementary factors.

We should remind the results of V.P. Potapov in the indicated direction. In the first place, every J-nonexpansive matrix possesses a J-polar representation, i.e. it may be represented in the form

$$T = UR, \tag{6}$$

where U is a J-unitary matrix, and R is a J-Hermitian matrix with nonnegative eigenvalues ([16], Chapter 2, Theorem 6).

In the second place, in the representation (6) the matrix R (the J-modulus of the J-nonexpansive matrix T) is determined uniquely. If, in addition, Det $T \neq 0$, then the matrix U is determined uniquely too ([16], Chapter 2, Theorem 8).

These results were extended by Ju.P. Ginzburg to operators in infinite dimensional spaces with an indefinite metric ([4], [5]). We shall give the necessary definitions.

An operator R is called J-selfadjoint, if $R^c = R$.

A J-selfadjoint operator $R \in B(\mathcal{H})$ is called the J-modulus of an operator $A \in B(\mathcal{H})$, if the following three conditions hold:

1) The spectrum of the operator R is nonnegative;

2) $R^2 = A^c A$;

3) Ker R = Ker $A^c A$.

Let V be a regular [2] closed J-isometric operator. An operator U is called partially J-isometric if

$$Uf = \begin{cases} Vf & \forall f \in \mathcal{D}(V) \\ 0 & \forall f \in \mathcal{D}(V)^{[\perp]}, \end{cases}$$

where the symbol $\mathcal{D}(V)^{[\perp]}$ denotes the J-orthogonal complement of the subspace $\mathcal{D}(V)$.

Finally, a representation

$$A = UR \tag{7}$$

of an operator $A \in B(\mathcal{H})$ is called a J-polar representation, if U is a partially J-isometric operator and R is the J-modulus of the operator A.

In the papers [4] and [5] of Ju.P. Ginzburg it was established that *every doubly-J-nonexpansive operator A possesses a J-polar representation (7), if in addition at least one of the next two conditions holds:*

1) *The operator A is completely invertible, i.e.* Ker $A = \{0\}$ *and* $\mathcal{R}(A) = \mathcal{H}$.

2) *The operator $A - I$ is compact*

Generalizing and deepening these results of Ju.P. Ginzburg, M.G. Krein and Ju.L. Shmul'jan in the paper [14] investigated the J-polar representation of a *strict plus-operator*, i.e. operator A of the form $A = \mu V$, where $\mu > 0$ and V is a *J-noncontractive operator*

$$\text{(i.e. } [Vx, Vx] \geq [x, x] \qquad \forall x \in \mathcal{H}).$$

It was proved that *the strict plus-operator A possesses the J-polar representation* (7) *if and only if the following conditions are fulfilled:*

    1) *The spectrum of operator $A^c A$ is nonnegative;*

    2) *The subspace $\mathcal{R}(A)$ is regular* [2].

If these conditions are fulfilled, then the operator R (the J-modulus of the operator A) is determined uniquely, and what is more, the operator U is determined uniquely if and only if at least one of the next equivalent conditions holds:

    a) Ker $A = \{0\}$;

    b) Ker $A^c = \{0\}$;

    c) The subspace Ker A^c is uniformly positive [2].

If, besides, the conditions 1) and 2) are realized, then the partially J-isometric operator U in the representation (7) may be chosen to be J-unitary (i.e. U is an isometric and $\mathcal{D}(U) = \mathcal{R}(U) = \mathcal{H}$) if and only if the two following conditions are fulfilled:

    α) The subspace Ker A^c is uniformly negative [2].

    β) dim Ker A^c = dim Ker A.

    At last, an important consequence of the above mentioned results by M.G. Krein and Ju.L. Shmul'jan is the existence of the J-polar representation (7) for every doubly strict plus-operator, in other words, for every operator which is collinear to a doubly-J-noncontractive operator [14].

    3. V.P. Potapov proved also that *every J-nonnegative matrix is J-unitary similar to a matrix having quasidiagonal shape* ([16], Chapter 2, Theorem 5). The generalization of this result to the infinite dimensional case of bounded J-nonnegative operators was obtained by Spitkovskii. A 2×2 block matrix $A = [A_{ij}]_{i,j=1}^2$ is called a *cross matrix* if $A_{22} = \overline{A}_{11}$ and $A_{22} = \overline{A}_{12}$. In the paper of Colpa [3] and with a shorter proof in the paper of Ikramov [6] the above mentioned result of V.P. Potapov (without reference to Potapov) was sharpened in the case of J-nonnegative matrices A, such that JA is a cross-matrix:

    a J-unitary matrix, realizing similarity, may be chosen as a cross-matrix.

This statement may be extended to the infinite dimensional case. We first give the necessary definitions, and then we formulate some results.

Let $\mathcal{H}^1$ be a Hilbert space with an orthonormal basis $\{e_\alpha\}_{\alpha \in \mathcal{U}}$. Put $\mathcal{H}^+ = \mathcal{H}^- = \mathcal{H}^1$ and form a new Hilbert space

$$\mathcal{H} = \mathcal{H}^+ \oplus \mathcal{H}^-,$$

in which the system $\{e_\alpha^+\}_{\alpha \in \mathcal{U}} \cup \{e_\alpha^-\}_{\alpha \in \mathcal{U}}$, where $e_\alpha^+ = e_\alpha \oplus 0$, $e_\alpha^- = 0 \oplus e_\alpha$, is an orthonormal basis. With the help of the operator J:

$$J(x \oplus y) = x \oplus (-y)$$

we introduce in $\mathcal{H}$ an indefinite metric $[u, v] = (Ju, v)$. The above introduced basis is also J-orthonormal

$$[e_\alpha^+, e_\beta^-] = \delta_{\alpha\beta}, \quad [e_\alpha^+, e_\beta^-] = 0, \quad [e_\alpha^-, e_\beta^-] = -\delta_{\alpha\beta}.$$

Let us introduce naturally the vector $\bar{x}$, the subspace $\bar{\mathcal{L}}$ and the operator $\bar{A}$, complex conjugate with respect to this basis to the vector x, the subspace $\mathcal{L}$ and the operator A correspondingly. An operator A is called *real* if $A = \bar{A}$.

Let us define the operator j:

$$j(x \oplus y) = y \oplus x$$

and call the operator A the cross-operator if $jAj = \bar{A}$. In particular, this concept in the case when A is defined on the whole space $\mathcal{H}$ coincides with a concept of a cross-matrix. Let B be a nonnegative selfadjoint cross-operator, such that the J-nonnegative operator $A = JB$ has a nonempty set $\rho(A)$ of regular points. We shall say, that the operator A has a *quasidiagonal shape with respect to the given basis*, if there exists a decomposition of the set $\mathcal{U} = \mathcal{U}_1 \cup \mathcal{U}_2 \cup \mathcal{U}_3$ in disjoint sets $\mathcal{U}_1, \mathcal{U}_2, \mathcal{U}_3$, such that the subspaces

$$\mathcal{H}_1^\pm = \text{c.l.s.}\{e_\alpha^\pm\}_{\alpha \in \mathcal{U}_1}, \quad \mathcal{H}_2^\pm = \text{c.l.s.}\{e_\alpha^\pm\}_{\alpha \in \mathcal{U}_2}, \quad \mathcal{H}_3 = \mathcal{H}_3^+ \oplus \mathcal{H}_3^-,$$

where $\mathcal{H}_3^\pm = \text{c.l.s.}\{e_\alpha^\pm\}_{\alpha \in \mathcal{U}_3}$, are invariant subspaces of the operator A, moreover, $A_1 = A \mid \mathcal{H}_1^+ = -jAj \mid \mathcal{H}_1^+$ is a positive real operator, $\mathcal{H}_2^\pm \subseteq \text{Ker } A$ and the operator $A \mid \mathcal{H}_3$ with respect to the mentioned decomposition of the space $\mathcal{H}_3$ has a matrix representation:

$$A \mid \mathcal{H}_3 = \begin{pmatrix} A_3 & J_3 A_3 \\ -J_3 A_3 & -A_3 \end{pmatrix},$$

where A_3 is a positive real operator, $J_3 A_3 = A_3 J_3$ and $J_3 e_\alpha^\pm = \pm e_\alpha^\pm$ ($\alpha \in \mathcal{U}_3$).

Let $\{E_\lambda\}$ be a J-spectral function of the J-nonnegative operator A and $\mathcal{L}_+ = \text{c.l.s.}\{(I - E_\lambda)\mathcal{H} \mid \lambda > 0\}$.

Theorem [1]. *In order that for the J-nonnegative operator $A = JB$ with $\rho(A) \neq \oslash$ and the cross-operator B there exists a J-unitary cross-operator U, such that the operator $U^{-1} A U$ has a quasidiagonal shape, it is necessary and sufficient that the subspace $\mathcal{L}_+$ is a Hilbert space with respect to the inner product $[u, v]$ $(u, v \in \mathcal{L}_+)$.*

In the definition above, the operator U, realizing similarity, is bounded. If this condition is omitted, then the criterion of reduction of the operator A to a quasidiagonal shape consists of pre-Hilbertness of the space $(\mathcal{L}_+, [,])$, and besides, in this case an operator U may be chosen, such that $A_3 = I$.

References

[1] T.Ja. Azizov: On diagonalization of nonnegative operator cross-matrices. X-V All-Union School on Operator Theory in Functional Spaces: Abstracts of the lectures. Uljanovsk, 1990 [In Russian].

[2] T.Ja. Azizov and I.S. Iohvidov: Linear Operators in spaces with an Indefinite Metric, J. Wiley, 1989; Translated from: The foundation of operator theory in spaces with an indefinite metric, M. Nauka, 1986 [In Russian]

[3] J.H.P. Colpa: Physica, Vol. A134, No. 2, 377-419.

[4] Ju.P. Ginzburg: On J-nonexpansive operator functions. Dokl. USSR 1957, Vol. 117, No 2, 171-173.

[5] Ju.P. Ginzburg: On J-nonexpansive operators in a Hilbert space. Nauchnye Zap. Fac. of Physics and Math, Odesskogo gosud Pedagogical Institute, 1958, Vol. 22, No. 1, 13-20. [In Russian]

[6] H.D. Ikramov: Journal Vychislitel'noj Mathematics and Mathematical Physics 1989, Vol. 29, No. 1, 3-14. [In Russian]

[7] E.I. Iohvidov: On a class of J-nonexpansive operators. Uspehi Math. Nauk. 1982, Vol. 37, No. 1, 127-128. [In Russian]

[8] E.I. Iohvidov: On linear operators which are J-nonexpansive together with their adjoints. Deponirovan V VINITI, 1983, No. 3284-83. [In Russian]

[9] E.I. Iohvidov: On symmetry of properties of a linear operator and its adjoint. Uspehi Math. Nauk. 1988, Vol.43, No. 5, 195-196. [In Russian]

[10] I.S. Iohvidov: Fractional linear functions of J-nonexpansive operators. Dokl. Acad. Nauk Armenian SSR, 1966, Vol. 42, No. 1, 3-8. [In Russian]

[11] I.S. Iohvidov: On Banch spaces with J-metric and classes of operators in these spaces. Isvestija Acad. Nauk Moldavian SSR, Series Physics, Technics and Mathematics, 1968 No. 1, 60-80. [In Russian]

[12] I.S. Iohvidov: On some classes of fractional-linear operator functions. Functional'nye prostranstva i operatornye uravnenija. Voronezh, 1970, pag. 18-44. [In Russian]

[13] M.G. Krein and Ju. L. Shmul'jan: On plus-operators in an indefinite metric space. Math. Issled. 1966, Vol.1, No. 1, 131-161. [In Russian]

[14] M.G. Krein and Ju. L. Shmul'jan: J-polar form of plus-operators. Math. Issled. 1966, Vol. 1, No. 2, 172-210. [In Russian]

[15] S.A. Kuzhel: J-nonexpansive operators. Theory of functions, functional analysis and its applications, Charkov, 1986, No. 45, 63-68. [In Russian]

[16] V.P. Potapov: Multiplicative structure of J-nonexpansive matrix functions. Trudy Mosk. Math. Ob. 1955, Vol. 4, 125-236. [In Russian]

[17] V.P. Potapov: Fractional-linear functions of matrices. Studies in the Theory of Operators and their Applications, Kiev, 1979, 75-97. [In Russian] Translated in: AMS Transl. Series 2, Vol 138, 21-35.

[18] V.S. Ritzner: A linear relation theory. Deponirovan V VINITI 1982, No. 846-82 [In Russian]

[19] I.M. Spitkovskii: On the block structure of J-unitary operators. Theory of functions, functional analysis and its applications, Charkov, 1978, No. 30, 129-138. [In Russian]

T.Ja.Azizov
P.O.Box 9
Voronezh-68
394068 Russia

Operator Theory:
Advances and Applications, Vol. 72
© 1994 Birkhäuser Verlag Basel

ON THE POTAPOV THEORY OF MULTIPLICATIVE

REPRESENTATIONS

Yu.P. Ginzburg and L.V. Shevchuk

The papers of V.P. Potapov on the multiplicative theory of analytic matrix functions published in the years 1950-1955, which contain the contents of his Doctor dissertation, have a special place in his creative scientific work: all the subsequent investigations carried out by him are connected with those publications and based on them.

The present paper is devoted to a survey of the fundamental results by Potapov on the theory of multiplicative representations as well as to some later publications of other authors in this direction.

I. As is known, in the disk $\mathbb{D} = \{\zeta \mid |\zeta| < 1\}$ any holomorphic function $x(\zeta) \not\equiv 0$ satisfying the inequality $|x(\zeta)| \leq 1$ ($\zeta \in \mathbb{D}$) admits the Blaschke-Riesz representation

$$x(\zeta) = e^{i\varphi_0} \prod_k \frac{\zeta_k - \zeta}{1 - \bar{\zeta}_k \zeta} \cdot \frac{|\zeta_k|}{\zeta_k} \cdot \exp \left\{ \int_0^{2\pi} \frac{\zeta + e^{i\vartheta}}{\zeta - e^{i\vartheta}} d\sigma(\vartheta) \right\}. \tag{1}$$

where ζ_k ($k = 1, 2, \dots$) are the zeroes of $x(\zeta)$ lying in $\mathbb{D}$ and repeated with the account of their order (for the simplicity of writing we assume that $x(0) \neq 0$). Furthermore, $\sigma(\vartheta)$ is a nondecreasing function on $[0, 2\pi]$, and φ_0 is a real constant.

Formula (1) clarifies the multiplicative structure of the function $x(\zeta)$; by the way it is natural to consider its exponential term as a continual product of factors

$$\exp \left\{ \frac{\zeta + e^{i\vartheta}}{\zeta - e^{i\vartheta}} d\sigma(\vartheta) \right\}$$

corresponding to the singularities of $x(\zeta)$ lying on the unit circle $\mathbb{T}$.

In the paper [1] V.P. Potapov generalized formula (1) for functions $X(\zeta)$, holomorphic in $\mathbb{D}$, the values of which are square matrices with norm ≤ 1, $\det X(\zeta) \not\equiv 0$. Bearing in mind further generalizations it is convenient for us to consider operators in the n-dimensional (complex) Hilbert space $\mathcal{H}^{(n)}$ instead of matrices; let us denote the corresponding set of operator functions by the symbol $\mathcal{K}^{(n)}$.

Following [1] we first consider the class $\mathcal{K}^{(n)}[\xi] \subset \mathcal{K}^{(n)}$ consisting of functions that are regular in the entire extended complex plane $\hat{\mathbb{C}}$, with the exception of the point $\xi \in \mathbb{T}$, and that are unitary valued on $\mathbb{T}\backslash\{\xi\}$. In the scalar case $(n = 1)$ this is the set of functions

$$U . \exp\left\{\frac{\zeta + \xi}{\zeta - \xi} A\right\} \qquad (A \geq 0, \ U^* U = I). \tag{2}$$

When $n > 1$ the class is wider and contains evidently products of factors of the type (2) (where A are Hermitian nonnegative operators), which are not reducible to (2) because of the noncommutivity of the operator multiplication, and it contains limits of such products. It turns out that the functions just described, in essence, exhaust the class $\mathcal{K}^{(n)}[\xi]$. Indeed, the following proposition holds.

THEOREM 1. *Any function X of the class $\mathcal{K}^{(n)}[\xi]$ can be represented in the form*

$$X(\zeta) = U \int_0^{\ell} \exp\left\{\frac{\zeta + \xi}{\zeta - \xi} dE(t)\right\}, \tag{3}$$

where U is a unitary operator in $\mathcal{H}^{(n)}$, $\ell \geq 0$, $E(t)$ is a Hermitian nondecreasing operator function on $[0, \ell]$, $\mathrm{tr}\, E(t) \equiv t$.

The integral in the right-hand side of (3) is a Stieltjes-Potapov multiplicative integral, that is, a limit for $\max(t_j - t_{j-1}) \to 0$ of integral products

$$\prod_{j=1}^n \exp\left\{\frac{\zeta + \xi}{\zeta - \xi}[E(t_j) - E(t_{j-1})]\right\} \qquad (0 = t_0 < t_1 < \cdots < t_{n-1} < t_n = \ell).$$

To describe the whole class $\mathcal{K}^{(n)}$ we note that an operator function of this class that is unitary on $\mathbb{T}$ and has a single singularity in $\hat{\mathbb{C}}$, a pole ξ of order 1, say, is of the form $B(\zeta) = U(\zeta P + Q)$ $(\xi = \infty)$

$$B(\zeta) = U\left(\frac{\alpha - \zeta}{1 - \overline{\alpha}\zeta} \cdot \frac{|\alpha|}{\alpha} P + Q\right) \qquad (\alpha = \overline{\xi}^{-1}, \ |\xi| > 1)$$

where U is a unitary operator, P is an orthogonal projection in $\mathcal{H}^{(n)}$, $Q = I - P$. Such a function is called an *elementary factor*. If $U = I$, then we shall call the elementary factor a *normalized* one.

THEOREM 2. *Each function X belonging to the class $\mathcal{K}^{(n)}$ can be represented in the form*

$$X(\zeta) = U \int_0^{\overset{\frown}{\ell}} \exp\left\{ \frac{\zeta + e^{i\vartheta(t)}}{\zeta - e^{i\vartheta(t)}}\, dE(t) \right\} \Pi(\zeta), \tag{4}$$

where U, ℓ, E are as in Theorem 1, $\Pi(\zeta)$ is either $\equiv I$ or is a (right) product (finite or convergent infinite) of normalized elementary factors $B_j(s)$†, $\vartheta(t)$ is a nondecreasing scalar function on $[0, \ell]$, and $0 \leq \vartheta(t) \leq 2\pi$.

During the late forties - early fifties there appeared a series of papers by M.S. Livšic [2-4] on the theory of operators different from normal ones. In this theory the main role is played by an operator function $W_A(\lambda)$, which corresponds to an operator A and is called the *characteristic function* of A (see, for example, formulas (6) and (9)). If the operator A is close (in certain sense) to a selfadjoint (unitary) operator, then $W_A(\lambda)$ and $W_A^{-1}(\lambda)$ are meromorphic in the upper half plane $\mathbb{C}_+$ (in $\mathbb{D}$), and $W(\lambda) = W_A(\lambda)$ satisfies in this domain the operator inequality

$$W^*(\lambda) J W(\lambda) \leq J \qquad (J^* = J, J^2 = I). \tag{5}$$

The function $W_A(\lambda)$ turns out to be nonexpansive $(J = I)$ if A is a dissipative operator (contraction).

It is important to note the following:

(1) from the equality of the characteristic functions of two completely nonselfadjoint (completely nonunitary) operators follows that these operators are unitary equivalent;

(2) there exists a close connection between invariant subspaces of the operator A and decompositions of its characteristic function $W_A(\lambda)$ into factors satisfying (5).

A notable advance in the study of connections of such a kind was realized by M.S. Livšic and V.P. Potapov [5].

Entire operator functions $W(\lambda)$ satisfying condition (5) in $\mathbb{C}_+$ and the condition $W^*(\lambda) J W(\lambda) = J$ $(\lambda \in \mathbb{R})$ appeared by this time in a number of papers ([6] et al.) on the spectral theory of differential systems.

---

† $\Pi(s) = \prod_{j=1}^{\overset{\frown}{r}} B_j(s) \ (r \leq \infty).$

The above papers [2-6] stimulated V.P. Potapov to publish the paper [1] and provided the motivation for the problem of extending the results of this paper (Theorems 1 and 2) to operator functions satisfying condition (5) in $\mathbb{D}$ or $\mathbb{C}_+$ (J-nonexpansive operator functions). A solution of this difficult problem has been obtained by V.P. Potapov and is outlined in detail in [7]. In order to formulate its main result we introduce the class $\mathcal{K}_J^{(n)} = \mathcal{K}_J^{(n)}(\mathbb{D})$ of J-nonexpansive operator functions, meromorphic in $\mathbb{D}$, with determinants not vanishing identically. Similarly, the class $\mathcal{K}_J^{(n)}(G)$ is defined, where G may be an arbitrary domain in $\hat{\mathbb{C}}$.

An elementary factor in the class $\mathcal{K}_J^{(n)} = \mathcal{K}_J^{(n)}(\mathbb{D})$ is a function $B \in \mathcal{K}_J^{(n)}$ satisfying (i) B has in $\hat{\mathbb{C}}$ a unique singularity ξ which is a pole of the first order; (ii) the operators $B(\zeta)$ ($\zeta \in \mathbb{T}\backslash\{\xi\}$) are J-unitary, that is, $B^*(\zeta)JB(\zeta) = B(\zeta)JB^*(\zeta) = J$. In [7] the general form of the elementary factors is clarified:

$$B(\zeta) = U(\frac{1-\bar{\xi}\zeta}{\xi-\zeta}\cdot\frac{\xi}{|\xi|}P+Q) \qquad (0 < |\xi| < \infty,\ |\xi| \neq 1),$$
$$B(\zeta) = U(\zeta^{-1}P+Q) \quad (\xi = 0) \qquad B(\zeta) = U(\zeta P+Q) \quad (\xi = \infty),$$
$$B(\zeta) = U(I + \frac{\zeta+\xi}{\zeta-\xi}E) \qquad (|\xi| = 1),$$

where U is a J-unitary operator, $P^2 = P$, $(1 - |\xi|)JP \leq 0$, $Q = I - P$, $JE \geq 0$, $E^2 = 0$. When $U = I$, we get normalized elementary factors.

THEOREM 3. *An arbitrary function X of the class $\mathcal{K}_J^{(n)}(\mathbb{D})$ can be represented in the form (4) where U is a J-unitary operator, $\Pi(\zeta)$ is a product of normalized elementary factors with poles $\xi \notin \mathbb{T}$ (possibly, $\Pi(\zeta) \equiv I$), $\ell \geq 0$, ϑ is a nondecreasing scalar function on $[0, \ell]$, $\vartheta(+0) = \vartheta(0)$, $0 \leq \vartheta(t - 0) = \vartheta(t) \leq 2\pi$ $(0 < t \leq \ell)$, $JE(t)$ is a Hermitian-nondecreasing operator function, tr $JE(t) \equiv t$.*

The proof of this proposition is considerably more complicated than that of Theorem 2 which concerns the definite case (i.e., $J = I$) and which is an immediate consequence of Theorem 3. The difficulties result, in particular, from the impossibility of using determinants, because it is easy to see that in $\mathcal{K}_J^{(n)}$ non-constant operator functions exist with constant determinants (for example, elementary factors with poles on $\mathbb{T}$). To overcome these difficulties V.P. Potapov used a number of subtle facts established by him. Let us mention only the linear-fractional transformations converting J-nonexpansive operators to nonexpansive ones and formulas connecting multiplicative integrals

$$Y(t) = \int_0^t \overset{\frown}{\exp}\{JdH(\tau)\}$$

($H(t)$ is a Hermitian nonincreasing operator function) and their J-modulus
$R(t) = \sqrt{JY^*(t)JY(t)}$ (the operator $JY^*(t)JY(t)$ turns out to have a positive spectrum).

To obtain from Theorem 3 the indefinite generalization of Theorem 1, which has important applications to inverse spectral problems, V.P. Potapov took advantage of the following proposition proved by him: if for a function $X \in \mathcal{K}_J^{(n)}$ the arc $\Gamma_{\alpha,\beta} = \exp\{i]\alpha, \beta[\}$ is a gap (that is, X is holomorphic and J-unitary on $\Gamma_{\alpha,\beta}$), then in the representation (4), with the conditions as in Theorem 3, we have $\vartheta([0,\ell]) \cap]\alpha, \beta[\neq \phi$.

Let us pay special attention to the subclass $\tilde{\mathcal{K}}_J^{(n)}$ of the class $\mathcal{K}_J^{(n)}$ consisting of rational J-unitary operator functions on $\mathbb{T}$.

THEOREM 4. If $X \in \tilde{\mathcal{K}}_J^{(n)}$, then X admits a representation in the form of a finite product of elementary factors.

Theorems 1-4 can be reformulated in an evident way for the classes $\mathcal{K}_J^{(n)}(\mathbb{C}_+)$, $\mathcal{K}_J^{(n)}(i\mathbb{C}_+)$ and for their naturally defined subclasses $\tilde{\mathcal{K}}_J^{(n)}(\mathbb{C}_+)$, $\tilde{\mathcal{K}}_J^{(n)}(i\mathbb{C}_+)$.

In the publications of V.P.Potapov and his pupils on the synthesis of electrical circuits (see survey [8]) the main role is played by a function of the class $\tilde{\mathcal{K}}_J^{(n)}(i\mathbb{C}_+)$ satisfying some conditions of symmetry (the simplest of these conditions is that of being real-valued). A clear physical meaning is contained in expansions of such functions into factors of the same type which have one, two or four poles.

It should be noted that the role played by the Schwarz lemma in proving the classical Blaschke theorem is replaced here by the following proposition (which helps in the proof of Theorem 3 too).

THEOREM 5. Let $X \in \mathcal{K}_J^{(n)}(\mathbb{D})$, $\zeta_1, \ldots, \zeta_k \ (\in \mathbb{D})$ be points of holomorphy of X, let $(\mathcal{H}^{(n)})^k$ be the Cartesian product of k copies of the space $\mathcal{H}^{(n)}$. Then for the operators in $(\mathcal{H}^{(n)})^k$ defined by the operator matrix

$$\mathcal{M}(X; \zeta_1, \ldots, \zeta_k) = \left((1 - \bar{\zeta}_r \zeta_s)^{-1} (J - X^*(\zeta_r) JX(\zeta_s)) \right)_{r,s=1}^k$$

the inequality $\mathcal{M}(X; \zeta_1, \ldots, \zeta_k) \geq 0$ is valid.

Inequalities of such a kind and their limit modifications are the main appara-
tus in the theory of Potapov on the matrix interpolation problems of R. Nevanlinna-Pick,
Schur, Carathéodory type, the moment problems (Hamburger and Stieltjes), the problems
of continuation of Hermitian-positive functions (M.G. Kreĭn). Under certain conditions
resolvent matrices† of these problems belong to the classes under consideration (for more
detail one is referred to the paper by T.S. Ivanchenko and L.A. Sahnovič in the present
volume).

2. Theorem 3 was used by M.S. Livšic in constructing a triangular model
acting in a Hilbert space $\mathcal{E}$ for a bounded linear operator A with an n-dimension $(n < \infty)$
imaginary component $A_{\Im} = (2i)^{-1}(A - A^*)$. Let us formulate this result for the operator
A with a real spectrum $\sigma(A)$. For this purpose let us consider the characteristic operator
function

$$W_A(\lambda) = I + 2i|A_{\Im}|^{\frac{1}{2}}(A^* - \lambda I)^{-1}|A_{\Im}|^{\frac{1}{2}}J \mid \mathcal{H}^{(n)} \tag{6}$$

$(\mathcal{H}^{(n)} = A_{\Im}\mathcal{E}, \quad J = \operatorname{sign} A_{\Im}|\mathcal{H}^{(n)})$. Since $W = W_A \in \mathcal{K}_J^{(n)}(\mathbb{C}_+)$, the following representation obtained from (4) holds

$$W(\lambda) = \int_0^{\overset{\frown}{\ell}} \exp\left\{i\frac{dH(t)}{\alpha(t) - \lambda}\right\}, \tag{7}$$

where $\alpha(t)$ is a nondecreasing scalar function on $[0, \ell]$, $JH(t)$ is a Hermitian nondecreasing
operator function, $\operatorname{tr} JH(t) \equiv t$. We construct from (7) the operator B acting in the
Lebesgue space L^2 of vector functions with values in $\mathcal{H}^{(n)}$:

$$(Bf)(x) = \alpha(x)f(x) + i\Pi(x)J\int_x^\ell \Pi(t)f(t)\,dt, \tag{8}$$

where $\Pi(x) = \sqrt{JH'(x)}$. The calculation shows that $W_B(\lambda) = W_A(\lambda)$ and therefore the
completely nonselfadjoint parts of the operators A and B are unitary equivalent.

To construct a triangular model of a bounded operator A under the less
heavy condition that $A_{\Im} \in \mathcal{S}_1$‡, M.S. Livšic [4] studied the multiplicative structure of

---

† The matrix of the coefficients of the linear-fractional transformation appearing in the formulas describing the general solutions of these problems.
   ‡ The class $\mathcal{S}_1$ consists of compact operators T acting in $\mathcal{H}$ and satisfying the
condition $\operatorname{tr}(T^*T)^{\frac{1}{2}} < \infty$.

infinitely dimensional operator functions which are characteristic functions of these operators. Thereby (by using a definition of a characteristic function, proposed by M.S. Brodskiĭ
[9,10], more general than (6), and for so-called operator nodes) the analogue of Theorem
3 was obtained for J-nonexpansive functions X, for which X and X^{-1} are meromorphic
in $\mathbb{D}$ and such that (1) $X(\zeta) - I \in S_1$ ($\zeta \in \mathbb{D}$), and (2) X has a gap on $\mathbb{T}$.

The last restriction was removed by Yu.P. Ginzburg [11,12] who considered
the set $\mathcal{K}_J^{S_1} = \mathcal{K}_J^{S_1}(\mathbb{D})$ of functions X that are holomorphic in $\mathbb{D}\backslash M_x$ (M_x is the set of
isolated points), such that the values of X are J-binonexpansive operators† on an separable
Hilbert space, and X satisfies the additional condition of existence of a point $\zeta_0 = \zeta_0(X)$
such that $X(\zeta_0)$ is boundedly invertible and $J - X^*(\zeta_0)JX(\zeta_0) \in S_1$.

THEOREM 6. [12] A function X belongs to the class $\mathcal{K}_J^{S_1}$ if and only if X
admits the representation (4) in which U, ℓ, ϑ, E satisfy the conditions of Theorem 3, and
$\Pi(\zeta)$ is a product of normalized elementary factors $B_j(\zeta)$ with poles $\zeta_j \notin \mathbb{T}$ and finite
dimensional projections P_j, for some $j_0 \in \mathcal{N}$

$$\Sigma_{j \geq j_0}|(1 - |\zeta_j|)\mathrm{tr}\ JP_j| < \infty \qquad \zeta_j = \infty \text{ for } 1 \leq j < j_0.$$

Under the formulated conditions, the limit processes in the right-hand side of formula (4)
converges with regard to the norm $\|A\|_1 = \mathrm{tr}\ (A^*A)^{\frac{1}{2}}$ ($A \in S_1$).

The proof of this theorem (as well as that of the multiplicative theorems
of V.P. Potapov and M.S. Livšic) is based on approximating the function X with finite
products of elementary factors.

V.T. Poljackiĭ has shown that a characteristic function [5]

$$\theta_T(\zeta) = V(T - \zeta J_{T^*}|I - TT^*|^{\frac{1}{2}}(I - \zeta T^*)^{-1}|I - T^*T|^{\frac{1}{2}}) \mid \mathcal{H}_T$$

$$(\mathcal{H}_T = \mathrm{clos}\ (I - T^*T)\mathcal{E}, \quad J_T = \mathrm{sign}\ (I - T^*T) \mid \mathcal{H}_T, \tag{9}$$

$$V : \mathcal{H}_{T^*} \to \mathcal{H}_T, \qquad V^*J_TV = J_{T^*}, \qquad VJ_{T^*}V^* = J_T)$$

of a linear bounded and boundedly invertible operator T acting on a $\mathcal{E}$ and satisfying the
condition $I - T^*T \in S_1$ belongs to the class $\mathcal{K}_J^{S_1}$ ($J = J_T$) and, therefore, admits the

---

† A linear bounded operator A is called J-binonexpansive if $A^*JA \leq J$ and
$AJA^* \leq J$. These inequalities are equivalent for the case when $A - I \in S_1$.

representation (4). Using this representation a triangular model of such an operator has been constructed in [13,14]. Similarly, Theorem 3 has been used by A.V. Kužel [15] for constructing a triangular model of a nonselfadjoint extension of an unbounded Hermitian operator with defect index (n, n).

L.A. Sahnovič [16-18] has established some conditions for the operator defined by formula (8) to be similar to a selfadjoint one, and has constructed the corresponding operator of similarity. In this matter, a proposition generalizing a classical theorem on boundary values of an integral of Cauchy type played an essential role. Let us formulate this proposition for the finite dimensional case and when $J = I$.

THEOREM 7. *Let the operator function $H(t)$ in (7) be Hermitian nondecreasing and absolutely continuous, $\alpha(t) \equiv t$. Then in almost all $x \in [0, \ell]$ there exist boundary values $W_\pm(x) = \lim_{y \to \pm 0} W(x + iy)$ which are determined by the formula*

$$W_\pm(x) = \lim_{\varepsilon \to +0} \int_0^{x-\varepsilon} \exp\left\{i\frac{dH(t)}{t-x}\right\} e^{\mp\pi H'(x)} \int_{x+\varepsilon}^{\ell} \exp\left\{i\frac{dH(t)}{t-x}\right\}.$$

3. During the late fifties - early sixties the papers by L.A. Sachnovič [19,20] and M.S. Brodskiĭ [21] were published; in these papers triangular representations were constructed without using multiplicative representations of characteristic operator functions. This made it possible to obtain multiplicative representations from the triangular ones. We now formulate the first result of such a kind which was deduced by M.S. Brodskiĭ [22,10] from the "abstract" triangular representation (which he derived too) of a Volterra operator A (i.e., for a compact operator A with all the spectrum in the point 0):

$$A = 2i \int_{\mathcal{R}} P A_\Im dP \tag{10}$$

($\mathcal{R} = \{P\}$ is a complete naturally ordered family of orthogonal projections on A-invariant subspaces; we shall not dwell on the precise definition of the integral (10)).

THEOREM 8. *Let $W(\lambda)$ be an entire function whose values are linear operators in Hilbert space $\mathcal{H}$, $J = J^* = J^{-1}$. If (i) $W(0) = I$, $W^*(\lambda)JW(\lambda) \le J$ ($\lambda \in \mathbb{C}_+$), $W^*(\lambda)JW(\lambda) = J$ ($\lambda \in \mathbb{R}$), and (ii) the operators $W(\lambda) - I$ are compact, then*

$$W(\lambda) = \int_0^{\ell} (I + i\lambda dE(t)), \tag{11}$$

where $\ell \geq 0$, $JE(t)$ is a Hermitian, strictly increasing $(E(t_1) \neq E(t_2)$ when $t_1 \neq t_2)$, strongly continuous function whose values are compact operators. The "binomial" multiplicative integral (11) is understood here in the sense of the convergence of a sequence of integral products in the uniform operator topology from a filter of partitionings of the segment $[0, \ell]$. When

$$\lim_{\lambda \to 0} \lambda^{-1}(W(\lambda) - I) \in \mathcal{S}_1,$$

then the integral (11) can be rewritten in the exponential form

$$W(\lambda) = \int_0^{\widehat{\ell}} \exp\{i\lambda dE(t)\} \qquad (E(\ell) \in \mathcal{S}_1). \tag{12}$$

Conversely, let $JE(t)$ be a Hermitian increasing strongly continuous function on $[0, \ell]$ with values that are compact operators, $E(0) = 0$. If $\sqrt{JE(\ell)}J\sqrt{JE(\ell)} \in \mathcal{S}_\omega$ †, then the integral (11) exists and is an entire function satisfying (i) and (ii). On the other hand, if $E(\ell) \in \mathcal{S}_1$, then the integral (12) exists and coincides in all λ with (11).

A further extension of the set of operators admitting triangular representations required considerable efforts and was realized in the publications [21-25]. Correspondingly, the set of operator functions decomposed multiplicatively was extended. Here, an essential role was played by the definition (in [26]) of a characteristic function that is more general than the one in formula (9). From the results of these investigations there seems to follow a binomial multiplicative representation of operator functions X of the class $\mathcal{K}_J^{\mathcal{S}_\omega}$ such that X and X^{-1} are both holomorphic in $\mathbb{D}$ (the definition of the class $\mathcal{K}_J^{\mathcal{S}_\omega}$ is similar to that of the class $\mathcal{K}_J^{\mathcal{S}_1}$). The theorem on such a multiplicative representation, which was proved by M.S. Brodskiĭ [22] under the additional assumption of the presence of a gap in the function X, is formulated in [25] without this restriction, and appears there as a consequence of a proposition, which, as was shown in [27], holds true only under some extra conditions. Nevertheless, there are good reasons to presume that the above assertion on the class $\mathcal{K}_J^{\mathcal{S}_\omega}$ holds true in general.

---

† The class $\mathcal{S}_\omega$ consists of all compact operators A acting in $\mathcal{H}$ and satisfying the condition $\sum k^{-1}\omega_k < \infty$, where (ω_k) is the nonincreasing sequence of eigenvalues of operator $(A^*A)^{\frac{1}{2}}$.

4. Among the operator functions to which the methods mentioned in part 3 apply there are such ones for which the operators $J - X^*(\zeta)JX(\zeta)$ are not compact. Examples of such operators are the operator functions that are characteristic functions of contractions which are similar to unitary operators [28], that is, nonexpansive operator functions X such that X and X^{-1} are both holomorphic in $\mathbb{D}$ and such that

$$\sup_{\zeta \in \mathbb{D}} \|X^{-1}(\zeta)\| < \infty. \tag{13}$$

The definition of a more general class $\mathcal{P}$ will be obtained if the latter condition is replaced by

$$\underline{\lim}_{\rho \uparrow 1} \int_0^{2\pi} \ln \|X^{-1}(\rho e^{i\vartheta})\| \, d\vartheta < \infty \tag{14}$$

and if one allows the function X^{-1} to have poles $\zeta_j \in \mathbb{D}$ such that $\sum(1 - |\zeta_j|) < \infty$ (each pole is repeated as many times as its order). The class $\mathcal{P}$ is the set of nonexpansive operator functions, holomorphic in $\mathbb{D}$, that possess a minorant [29] (a scalar multiple [30]), that is, a scalar holomorphic function $m(\zeta) \not\equiv 0$ such that $X^*(\zeta)X(\zeta) \geq |m(\zeta)|^2 I$. Among the minorant functions $X \in \mathcal{P}$ there exists the best minorant $m_X(\zeta)$, which is uniquely determined up to a constant factor ε, $|\varepsilon| = 1$. It is easy to see that the function $d_X(\zeta) = \det VX(\zeta)$ is a minorant (not necessarily the best) of the function $X \in \mathcal{K}_I^{S_1}$ (here V is a unitary operator for which $VX(\zeta) - I \in S_1$; the existence of such a V, which follows directly from Theorem 6, can also be easily proved without using Theorem 6).

The methods used in the proof of the first assertion of the next theorem are different from the approximation methods of Potapov and are not based on the theory of triangular representations; here well-known theorems of Beurling on shift operators and of Devinatz on factorization of Hermitian positive operator functions on $\mathbb{T}$ are used. The arguments for the assertions on uniqueness come essentially from the theorem of Potapov (mentioned in part 1) on the modulus of a multiplicative integral [7, 31].

THEOREM 9. [32, 33]. *The operator function* $X(\zeta)$ ($\zeta \in \mathbb{D}$) *belongs to the class* $\mathcal{P}$ *if and only if* X *admits the representation* (4) *in which* (i) U *is a unitary operator,* (ii) $\Pi(\zeta) \equiv I$ *or is a convergent product of elementary factors*†

$$B_j(\zeta) = \frac{\zeta_j - \zeta}{1 - \bar{\zeta}_j \zeta} \cdot \frac{|\zeta_j|}{\zeta_j} P_j + Q_j,$$

---

† For simplicity we consider the case when $X(0)$ is boundedly invertible.

where P_j are orthogonal projections in $\mathcal{H}$, $Q_j = I - P_j$, $\sum(1 - |\zeta_j|) < \infty$, and (iii) $\vartheta(t)$ is a nondecreasing scalar function on $[0, \ell]$, $\vartheta(+0) = \vartheta(0)$, $\vartheta(t - 0) = \vartheta(t)$ $(0 < t \leq \ell)$, $0 \leq \vartheta(t) < 2\pi$ for $0 \leq t < \ell$; $E(t)$ is a Hermitian increasing continuous operator function on $[0, \ell]$, $E(0) = 0$,

$$\mathrm{var}_{[0,\ell]} E = \sup \Sigma_j \|E(t_j) - E(t_{j-1})\| < \infty \qquad (0 = t_0 < t_1 < \ldots < t_{n-1} < t_n = \ell).$$

Under the formulated conditions the limit processes in the right-hand side of formula (4) converge with regard to the uniform operator norm. In this representation the $U\Pi(\zeta)$ is defined uniquely, the function $\vartheta(t)$ is defined uniquely up to a strictly increasing continuous substitution of the variable t; after choosing the function ϑ for each of its points of growth‡ t_0 the operators $E(t_0)$ and

$$X_{t_0}(\zeta) = \int_0^{t_0} \exp\left\{\frac{\zeta + e^{i\vartheta(t)}}{\zeta - e^{i\vartheta(t)}} dE(t)\right\} \qquad (\zeta \in \mathbb{D}). \tag{15}$$

are also defined uniquely by X.

The matrix function example

$$X(\zeta) = \exp\left\{\frac{\zeta + 1}{\zeta - 1}\begin{pmatrix} 1 & 0 \\ 0 & 1 \end{pmatrix}\right\}$$

shows that for points t_0 which belong to an interval of constancy of the function ϑ the operator $E(t_0)$ is not uniquely defined, generally speaking.

In a common case, the following result of M.S. Brodskiĭ and G.E. Kisilevskiĭ [10] (which is closely related to the criterion of these authors on the unicellular structure of dissipative Volterra operators) allows one to answer the question about conditions of uniqueness of $E(t_0)$ which are determined from $X \in \mathcal{K}_I^{S_1}$.

THEOREM 10. *Let the operator function $W(\lambda)$ have the representation (12) satisfying the conditions of Theorem 8 when $J = I$ and the normalized condition $\mathrm{tr}\ E(t) \equiv t$. This representation is unique if and only if the exponential type of the function $W(\lambda)$ is equal to ℓ (that is, equal to the type of the function $\det W(\lambda)$).*

---

‡ That is, the points which don't belong to any (open) interval of constancy of the function ϑ.

From Theorem 10 and Theorem 9 the following proposition may be derived.

THEOREM 11. [32] *Let* $X \in \mathcal{K}_I^{S_1}$, *and assume that* $X^{-1}(\zeta)$ *exist for all* $\zeta \in \mathbb{D}$. *Let*

$$m_X(\zeta) = \exp\left\{ \int_0^{2\pi} \frac{\zeta + e^{i\varphi}}{\zeta - e^{i\varphi}} d\sigma(\varphi) \right\}, \qquad d_X(\zeta) = \exp\left\{ \int_0^{2\pi} \frac{\zeta + e^{i\varphi}}{\zeta - e^{i\varphi}} d\omega(\varphi) \right\}, \qquad (16)$$

where σ *and* ω *are nondecreasing functions on* $[0, 2\pi]$ *which are continuous from the left. For* X *to admit a unique representation* $X(\zeta) = UX_\ell(\zeta)$ *(see (15)), in which the operator function* E *is normalized by the condition* $\mathrm{tr}E(t) \equiv t$, *it is necessary and sufficient that the function* $\omega(\varphi) - \sigma(\varphi)$ *is continuous on* $[0, 2\pi]$. *Each of the functions* σ, ω *is absolutely continuous with respect to the other.*

Because of boundedness, the operator function $X \in \mathcal{P}$ has in almost all points $e^{i\varphi} \in \mathbb{T}$ a strong radial boundary value $X(e^{i\varphi})$. If for almost all $\varphi \in [0, 2\pi]$ the operator $X(e^{i\varphi})$ is unitary, then X is called an *inner function* ($X \in \mathcal{P}^{(i)}$). The function $X \in \mathcal{P}$ is called *outer* ($X \in \mathcal{P}^{(e)}$) if X has no nontrivial left inner divisors in the class $\mathcal{P}$. One has $X \in \mathcal{P}^{(i)}$ ($X \in \mathcal{P}^{(e)}$) if and only if m_X is an inner (outer) function. In this proposition when $X \in \mathcal{K}_I^{S_1}$ it is possible to substitute m_X for d_X in the latter statement, which follows from the last assertion of Theorem 11.

THEOREM 12. *In order that* $X \in \mathcal{P}$ *it is necessary and sufficient that the following representation holds:*

$$X(\zeta) = U\Pi(\zeta)X^{(p)}(\zeta)X^{(s)}(\zeta)X^{(a)}(\zeta), \qquad (17)$$

where U *and* Π *are as described in Theorem 9,*

$$X^{(p)}(\zeta) = \int_0^{\ell} \exp\left\{ \frac{\zeta + e^{i\vartheta(t)}}{\zeta - e^{i\vartheta(t)}} dG(t) \right\}, \qquad (18)$$

$$X^{(s)}(\zeta) = \int_0^{2\pi} \exp\left\{ \frac{\zeta + e^{i\varphi}}{\zeta - e^{i\varphi}} d\Sigma_s(\varphi) \right\}, \qquad (19)$$

$$X^{(a)}(\zeta) = \int_0^{2\pi} \exp\left\{ \frac{\zeta + e^{i\varphi}}{\zeta - e^{i\varphi}} d\Sigma_a(\varphi) \right\}. \qquad (20)$$

(Here ϑ *is a function satisfying the conditions (iii) of Theorem 9 of which the sum of the lengths of its intervals of constancy is equal to* ℓ, $G(t)$ *is a Hermitian increasing function*

on $[0, \ell]$, $\mathrm{var}_{[0;t]}G \equiv t$, Σ_s and Σ_a are Hermitian nondecreasing continuous singular† and absolutely continuous‡ operator functions, $\Sigma_s(0) = \Sigma_a(0) = 0$). The functions $U\Pi$, $X^{(p)}$, Σ_s are defined uniquely by X, and Σ_a by the function $X^*(e^{i\varphi})X(e^{i\varphi})$ ($\varphi \in [0, 2\pi]$). We have $X \in \mathcal{P}^{(i)}$ ($X \in \mathcal{P}^{(e)}$) if and only if $X'^{(a)}(\zeta) \equiv I$ ($\Pi(\zeta) \equiv X^{(p)}(\zeta) \equiv X^{(s)}(\zeta) \equiv I$) in (17).

In the papers [34-37] representations of the type (17) - (20) are given for matrix and operator analogues of scalar functions in the classes $\mathcal{N}$ (R. Nevanlinna), $\mathcal{D} = \mathcal{N}_+$ (V.I. Smirnov), H^δ (G. Hardy). So, for example, in order that the elements of the matrix function $X(\zeta)$ belong to the class H^δ ($0 < \delta < \infty$) and $\det X(\zeta) \not\equiv 0$ it is necessary and sufficient to have the representation (17) - (20) in which $\mathcal{U}, \Pi, \vartheta, G, \Sigma_s$ satisfy the conditions of Theorem 12, and Σ_a is a Hermitian-valued absolutely continuous matrix function such that

$$\int_0^{2\pi} \exp\{-2\pi\delta\mu(\varphi)\}\, d\varphi < \infty,$$

$\mu(\varphi)$ is the least eigenvalue of the matrix $\Sigma_a'(\varphi)$. The last two assertions of Theorem 12 are transferred to the above classes.

In the papers [38-40] Theorems 9 and 12 are used for studying operators whose characteristic functions belong to the class $\mathcal{P}$.

5. Unfortunately, several problems that are solved for nonexpansive operator functions remain open for the much more complicated and important indefinite case. Let us consider the results obtained so far that are the most interesting ones (from our point of view) and that are at the same time simple to formulate.

Since each operator function $X \in \mathcal{K}_J^{S_1}$ has almost everywhere strong radial limited values $X(\xi)$ ($\xi \in \mathbb{T}$) [41], J-inner and J-outer functions of this class may be determined naturally. The fact that a convergent Blaschke-Potapov product in case of the class $\mathcal{K}_J^{(n)}$ is a J-inner function was proved in [42]. In the paper [43] this fact was deduced from the following general proposition

THEOREM 13. *Let* $X_k(\xi)$ ($k \in \mathbb{N}$), *and* $X(\zeta)$ *be functions of the class* $\mathcal{K}_J^{(n)}$

---

† That is, a weak derivative $\Sigma_s'(\varphi) = 0$ is almost everywhere on $[0, 2\pi]$.
‡ With regard to the uniform operator norm.

$(n < \infty)$ for which $\lim X_k(\zeta) = X(\zeta)$, $\det X(z) \not\equiv 0$, $X^*(\zeta)JX(\zeta) \le X_k^*(\zeta)JX_k(\zeta)$ $(\zeta \in \mathbb{D})$. Then a subsequence X_{k_j} exists such that in almost all $\xi \in \mathbb{T}$

$$\lim_{j \to \infty} X_{k_j}(\xi) = X(\xi).$$

A J-inner function $X \in \mathcal{K}_J^{(n)}$ is called *singular* if all matrix elements of X and X^{-1} belong to the class $\mathcal{N}_+$, and *regular* if it does not have any nontrivial singular right divisors [44].

Let us consider orthogonal projections $P_\pm = \frac{1}{2}(I \pm J)$ and subspaces $\mathcal{H}_\pm = P_\pm \mathcal{H}^{(n)}$. Let

$$\begin{pmatrix} X_{11}(\zeta) & X_{12}(\zeta) \\ X_{21}(\zeta) & X_{22}(\zeta) \end{pmatrix} \tag{21}$$

be the matrix of the operator function $X \in \mathcal{K}_J^{(n)}$ corresponding to the decomposition $\mathcal{H}^{(n)} = \mathcal{H}_+ \oplus \mathcal{H}_-$. When proving the following proposition it is used that $X \in \mathcal{K}_J^{(n)}$ $(J \neq \pm I)$ is a left product of elementary factors B_j with poles $\xi_j \notin \mathbb{T}$ $(X = \widehat{\prod_j} B_j)$ if and only if (21) is a solution matrix for the so-called bitangential interpolation problem of Schur-Nevanlinna-Pick.

THEOREM 14. [44] *For $X \in \mathcal{K}_J^{(n)}$ to be a left product of elementary factors with poles not lying on $\mathbb{T}$ it is necessary and sufficient for X to be a regular operator function and*

$$\lim_{\rho \uparrow 1} \int_0^{2\pi} \ln |\det X_{22}(\rho e^{i\varphi})| \, d\varphi = \int_0^{2\pi} \ln |\det X_{22}(e^{i\varphi})| \, d\varphi$$

$$\lim_{\rho \downarrow 1} \int_0^{2\pi} \ln |\det X_{11}(\rho e^{i\varphi})| \, d\varphi = \int_0^{2\pi} \ln |\det X_{11}(e^{i\varphi})| \, d\varphi$$

(Here we suppose that the function X is extended on $\mathbb{D}_+ = \{\zeta \mid |\zeta| > 1\}$ by the formula $X(\zeta) = JX^{-1*}(\overline{\zeta}^{-1})J$).

In the particular case $J = I$ a corresponding result (which implies that the condition " $\det X(\zeta)$ is a Blaschke product" is satisfied) can be easily proved directly, even for the class $\mathcal{K}_I^{S_1}$: this proposition was already used in the paper by M.S. Livšic [4], which proves the criterion of completeness of a completely nonselfadjoint operator A, $0 \le A_J \in \mathcal{S}_1$ (see also [13,15]).

In connection with Theorem 14 it is worthwile to mention [44] that a convergent product of elementary factors with poles on $\mathbb{T}$ is a singular operator function.

Despite the fact that the order of the sequence of singularities in the Blaschke-Potapov product can be chosen arbitrarily, such a proposition holds which was proved in a finite-dimensional case by Potapov [7].

THEOREM 15. [12] Let $X \in \mathcal{K}_J^{S_1}$, Π_k $(k = 1, 2)$ be convergent right products of elementary factors $B_{k,j}$ with poles not lying on $\mathbb{T}$ $(\Pi_k = \widehat{\prod_j} B_{k,j})$ such that the functions $X_1 = X\Pi_1^{-1}$, $X_2 = X\Pi_2^{-1}$ belong to $\mathcal{K}_J^{S_1}$ and these functions together with X_1^{-1} and X_2^{-1} are holomorphic in $\mathbb{D}$. Then $\Pi_2(\zeta) = U\Pi_1(\zeta)$ $(\zeta \in \mathbb{D})$ for some J-unitary operator U.

Of course, this result carries over to left products.

It can be easily seen that in the class $\mathcal{P} \supset \mathcal{K}_I^{S_1}$ each left (right) Blaschke-Potapov product is right (left). For an arbitrary J this is not the case, as was shown by V.E. Katsnelson.

THEOREM 16. [45] Let $J \neq \pm I$. Given an arbitrary singular J-inner function $Y \in \mathcal{K}_J^{(n)}$, there exists a right product Π of elementary factors B_j with poles $\zeta_j \notin \mathbb{T} \cup \mathbb{D}$ such that $X = Y\Pi$ is a left Blaschke-Potapov product. Conversely, if Π_1 and Π_2 are left and right Blaschke-Potapov products with poles $\notin \mathbb{T} \cup \mathbb{D}$ and $\Pi_2 = Y\Pi_1$, where Y $(\in \mathcal{K}_J^{(n)})$ and Y^{-1} are holomorphic in $\mathbb{D}$, then Y is a singular J-inner matrix function.

A number of results on uniqueness of multiplicative integral expansions of the functions $X \in \mathcal{K}_J^{S_1}$ is obtained by L.M. Zemskov [46] under some restrictions superposed on the behaviour X near the point $\zeta = 1$. For simplicity, in the subsequent formulations we shall confine ourselves to the finite-dimensional case.

THEOREM 17. [46] Let $X \in \mathcal{K}_J^{(n)}$ together with X^{-1} be holomorphic in $\mathbb{D}$, and assume that for any sequence $\{\zeta_j\} \subset \mathbb{D}$ converging nontangentially to 1 we have

$$\overline{\lim_{j \to \infty}} \|(X^*(\zeta_j)JX(\zeta_j)^{\pm 1}\| < \infty. \tag{22}$$

Then in the representation (4) (in case $\Pi(\zeta) \equiv I$) which satisfies the condition of Theorem 3, the number ℓ, the function ϑ, and operators $E(t_0), X_{t_0}(\zeta)$ (15) for each point t_0 of the

growth of ϑ are determined uniquely†.

L.A. Sahnovič [47], taking advantage of his method of operator identities, investigated a wide class of multiplicative representations. In particular, from his results it follows directly that each $X \in \mathcal{K}_J^{(n)}$ for which

$$\sup_{\zeta \in \mathbb{D}} \| X^{\pm 1}(\zeta) \| < \infty \qquad (23)$$

has the representation

$$X(\zeta) = U \int_0^{2\pi} \exp \left\{ \frac{\zeta + e^{i\varphi}}{\zeta - e^{i\varphi}} \right\} dG(\varphi),$$

where U is a J-unitary operator, $JG(\varphi)$ is a continuous Hermitian nondecreasing operator function ($G(0) = 0$) which is constructed from X by means of an effective algorithm. It follows from Theorem 17 that this algorithm leads to a single function G possessing the above properties.

It should be noted that if for the characteristic function X of the operator T condition (23) is satisfied, then T is linearly similar to a unitary operator [48].

The condition (22), in particular, is satisfied for those $X \in \mathcal{K}_J^{(n)}$ for which the following limit exists

$$\lim_{\rho \uparrow 1} (1 - \rho)^{-1} (J - X^*(\rho) J X(\rho)).$$

As was found by L.V. Shevchuk [49], these functions are those and only those which are characteristic functions of operators $T = (A - z_0 I)(A + z_0 I)^{-1}$ where A is the sum of a selfadjoint (bounded or unbounded) and a finite-dimensional operator, $z_0, \bar{z}_0 \notin \sigma(A)$, $z_0 \in \mathbb{C}_+$. Multiplicative representations of such operator functions are studied in [50,51].

The effective formula for recovering the J-outer function $X \in \mathcal{K}_J^{(2)}$ ($J \neq \pm I$) from the J-modulus of its limited values was obtained by P.M. Yudickiĭ [52,53] who reduced

---

† The question on the uniqueness of $E(t_0)$ for points t_0 which belong to an interval of constancy of ϑ implies here as well as in case $J = I$ the investigation of integrals of the form (12). Many numerical results of this aspect are not considered by us; they are discussed in the paper by L.A. Sahnovič in this volume.

this problem to a generalized interpolation problem for which the matrix (21) is the solution (in fact, this formula remains true in any class $\mathcal{K}_J^{(n)}$, $n < \infty$). It should be noted that getting multiplicative representations of J-inner and J-outer operator functions is a rather timely problem, which as far as we know, is not yet solved.

REFERENCES

1. V.P. Potapov, On holomorphic matrix functions bounded in the unit circle, *Dokl. Akad. Nauk SSSR* 72 (1950), 849-852.

2. M.S. Livšic, On a class of linear operators in Hilbert space, *Mat. Sb* 9 (61) (1946), 299-262.

3. M.S. Livšic, Isometric operators with equal defect index, quasiunitary operators, *Mat. Sb* 26 (68) (1950), 247-264.

4. M.S. Livšic, On spectral resolution of nonselfadjoint linear operators, *Mat. Sb* 34 (76) (1954), 145-199; English Transl., *Amer. Math. Soc. Transl. (Series 2)* 5 (1957), 67-114.

5. M.S. Livšic and V.P. Potapov, Multiplication theorem on the characteristic matrix functions, *Dokl. Akad. Nauk SSSR* 62 (1950), no. 4, 625-628.

6. M.G. Kreĭn, On the theory of entire matrix functions of exponential type, *Ukrain. Mat.Žurn.* 3 (1951), no. 2, 164-173.

7. V.P. Potapov, The multiplicative structure of J-nonexpansive matrix functions, *Trudy Moskov. Mat. Obšč.* 4 (1955), 125-236; English Transl., *Amer. Math. Soc. Transl. (Series 2)* 15 (1960), 131-243.

8. A.V. Efimov and V.P. Potapov, J-contractive matrix functions and their role in the theory of electric chains, *Uspehi Mat. Nauk* 28 (1973), no. 1, 65-130.

9. M.S. Brodskiĭ and M.S. Livšic, Spectral analysis of nonselfadjoint operators and intermediate systems, *Uspehi Mat. Nauk* 13 (1958), no. 1, 3-85; English Transl., *Amer. Math. Soc. Transl. (Series 2)* 13 (1960), 265-346.

10. M.S. Brodskiĭ, *Triangular and Jordan representations of linear operators*, "Nauka", Moscow, 1969; English Transl., Amer. Math. Soc. Transl., Vol. 32, Providence, R.I. 5 1971.

11. Yu.P. Ginzburg, On J-nonexpansive operator functions, *Dokl. Akad. Nauk SSSR* 117 (1957), 171-173.

12. Yu.P. Ginzburg, On multiplicative representations of J-nonexpansive operator functions, *Mat. Issled.* 2 (1967), no. 2, 52-83; no.3, 20-51; English Transl., *Amer. Math. Soc. Transl. (Series 2)* 96 (1970), 189-221.

13. V.T. Poljackiĭ, On the reduction of quasiunitary operators to a triangular form, *Dokl. Akad. Nauk SSSR* 113 (1957), 756-759.

14. V.T. Poljackiĭ, On the reduction of operators of class K to a triangular form, *Naučn. Zap. Odessk. Gos. Ped. Inst.* 24 (1959), no. 1, 13-15.

15. A.V. Kužel, Spectral analysis of unbounded nonselfadjoint operators, *Dokl. Akad. Nauk SSSR* 125 (1959), 35-37.

16. L.A. Sahnovič, On the reduction of nonselfadjoint operators to a diagonal form, *Dokl. Akad. Nauk SSSR* 115 (1957), no. 3, 462-465.

17. L.A. Sahnovič, Limiting values of a multiplicative integral, *Ukrain. Mat. Žurn.* 11 (1959), no. 3, 275-286.

18. L.A. Sahnovič, Dissipative operators with an absolutely continuous spectrum, *Trudy Moskov. Mat. Obšč.* 19 (1986), 211-270.

19. L.A. Sahnovič, On the reduction of nonselfadjoint operators to a triangular form, *Izv. Vysš. Učebn. Zaved., Mathematika* (1959), no. 1, 180-186.

20. L.A. Sahnovič, The research of a triangular form of nonselfadjoint operators, *Izv. Vysš. Učebn. Zaved., Mathematika* (1959), no. 4, 141-149.

21. M.S. Brodskiĭ, Triangular representations of some operators with completely continuous imaginary part, *Dokl. Akad. Nauk SSSR* 133 (1960), 1271-1274; English Transl., *Soviet Math. Dokl.* 1 (1960).

22. M.S. Brodskiĭ, Miltiplicative representations of certain analytic operators functions, *Dokl. Akad. Nauk SSSR* 138 (1961), 751-754; English Transl., *Soviet Math. Dokl.* 2 (1961), 695-698.

23. I.C. Gohberg and M.G. Kreĭn, On the multiplicative representations of the characteristic functions of operators close to the unitary ones, *Dokl. Akad. Nauk SSSR* 164 (1965), 732-735; English Transl., *Soviet Math. Dokl.* 6 (1965), 1279-1283.

24. V.M. Brodskiĭ, Multiplicative representations of the characteristic functions of contraction operators, *Dokl. Akad. Nauk SSSR* 173 (1967), 256-259; English Transl., *Soviet Math. Dokl.* 8 (1967), 362-366.

25. V.M. Brodskiĭ, I.C. Gohberg and M.C. Kreĭn, General theory on triangular representations of linear operators and multiplicative representations of their characteristic funtions, *Funkcional. Anal. i Priložen.* 3 (1969), no. 4, 1-27.

26. V.M. Brodskiĭ, I.C. Gohberg and M.C. Kreĭn, Definition and main properties of a characteristic funtion of a *J*-node, *Funkcional. Anal. i Priložen.* 4 (1970), no. 1, 88-90.

27. N.I. Baranov and M.S. Brodskiĭ, Triangular representations of operators with unitary spectrum, *Funkcional. Anal. i Priložen.* 16 (1982), no. 1, 58-59.

28. I.C. Gohberg and M.C. Kreĭn, On a description of contraction operators which are similar to unitary, *Funkcional. Anal. i Priložen.* 1 (1967), no. 1, 38-60.

29. Yu.P. Ginzburg, On divisors and minorants of operator functions of bounded form, *Mat. Issled., Kishinev*, 2 (1967), no. 4, 47-72.

30. B.Sz.-Nagy and C. Foias, *Harmonic analysis of operators on Hilbert space*, Amsterdam, London, Budapest (1970).

31. V.P. Potapov, The theorem on modulus, *Teor. Funkciĭ Funkcional. Anal. i Priložen.* 38 (1982), 91-101; 39 (1983), 95-106.

32. Yu.P. Ginzburg, Multiplicative representations and minorants of bounded analytic operator functions, *Funkcional. Anal. i Priložen.* 1 (1967), no. 3, 9-23.

33. Yu.P. Ginzburg and L.M. Zemskov, On the multiplicative representations of operator functions of bounded form, *Teor. Funkciĭ, Funkcional. Anal. i Priložen.* 53 (1990), 108-118.

34. Yu.P. Ginzburg, The factorization of analytic matrix functions, *Dokl. Akad. Nauk SSSR* 159 (1964), 489-492; English Transl., *Soviet Math. Dokl.* 5 (1961), 1510-1514.

35. Yu.P. Ginzburg, Multiplicative representations of bounded analytic operator functions, *Dokl. Akad. Nauk SSSR* 170 (1966), 23-26; English Transl., *Soviet Math. Dokl.* 7 (1966), 1125-1128.

36. Yu.P. Ginzburg, Multiplicative representations of operator functions of bounded form, *Uspehi Mat. Nauk* 22 (1967), no. 1, 163-165.

37. Yu.P. Ginzburg, On reconstruction of the multiplicative integral from its modulus, *Teor. Funkciĭ, Funkcional. Anal. i Priložen.* 41 (1984), 135-143.

38. Yu.P. Ginzburg and R.L Mogilevskaja, A certain class of dissipative operators with slowly increasing resolvent, *Funkcional. Anal. i Priložen.* 3 (1969), no. 4, 83-84.

39. Yu.P. Ginzburg and R.L. Mogilevskaja, The spectral functions of contractions with slowly increasing resolvent, *Dokl. Akad. Nauk SSSR* 207 (1972), 517-520.

40. R.L Mogilevskaja, The triangular models of the operators with slowly increasing resolvent, *Teor. Funkciĭ, Funkcional. Anal. i Priložen.* 27 (1977), 95-108.

41. V.F. Veselov and S.N. Naboko, Determinant of characteristic function and singular spectrum of nonselfadjoint operator, *Mat. Sb.* 129 (1986), no. 1, 20-39.

42 D.Z. Arov and L.A. Simakova, On boundary values of converging sequence of J-contractive matrix functions, *Mat. Zam.* 19 (1976), no. 4, 491-500.

43 D.Z. Arov, On boundary values of converging sequence of meromorphic matrix functions, *Mat. Zam.* 25 (1979), no. 3, 335-340.

44. D.Z. Arov, γ-deriving matrices, J-inner matrix functions and related extrapolation matrix functions problems, I, II, III, *Teor. Funkciĭ, Funkcional. Anal. i Priložen.* 51 (1989); 52 (1989); 53 (1990).

45. V.E. Katsnelson, Left and right Blaschke-Potapov products and Arov-singular matrix valued functions, *Integr. Equat. and Oper. Theory* 13 (1990), no. 6, 836-848.

46. L.M. Zemskov, The uniqueness of multiplicative representations of analytic J-nonexpansive operator functions, deposited in *Ukr. NIINTI*, no. 2961 Uk 87 Dep., 27 pages, 1987.

47. L.A. Sahnovič, The factorization problem and operatioral identities, *Uspehi Mat. Nauk* 41 (1986), no. 1, 3-55.

48. L.A. Sahnovič, Nonuitary operators with absolutely continuous spectrum on the unit circle, *Dokl. Akad. Nauk SSSR*, 181 (1968), no. 3.

49. L.V. Shevchuk, Boundary behaviour of characteristic functions and the trace formula for dissipative operators, *Dokl. Akad. Nauk Ukr. SSR*, Ser."A", 7 (1986), 21-24.

50. Yu.P. Ginzburg and L.V. Shevchuk, On analytic J-nonexpansive operator functions satisfying J̇ulia condition, deposited in *Ukr. NIINTI*, no. 274 Uk 86, 56 pages, 1986.

51. L.V. Shevchuk, On multiplicative representations of J-nonexpansive operator functions satisfying J̇ulia condition, deposited in *Ukr. NIINTI*, no. 1315 Uk 86, 41 pages, 1986.

52. P.M. Yudickiĭ, On reconstitution of the J-contractive analytic matrix function from boundary values of its J-form and connected with this "interpolation" problem, *Teor. Funkciĭ, Funkcional. Anal. i Priložen.* 44 (1985), 141-143.

53. P.M. Yudickiĭ, Outer-inner factorization of *J*-expansive invertible matrix functions *Teor. Funkciĭ, Funkcional. Anal. i Priložen.* 46 (1986), 132-135.

Odessa Technological

Institute of Food Industry,

Sverdlov Str., 112, Odessa,

Ukraine 270039

48

Operator Theory:
Advances and Applications, Vol. 72
© 1994 Birkhäuser Verlag Basel

AN OPERATOR APPROACH TO THE POTAPOV SCHEME
FOR THE SOLUTION OF INTERPOLATION PROBLEMS

T. S. Ivanchenko and L. A. Sakhnovich

Interpolation problems play a significant role both in applied and theoretical investigations. Classical interpolation problems include those of Nevanlinna-Pick, Caratheodory, Schur, and Hamburger, as well as the problem of trigonometric moments.

Recently new interesting approaches [1] - [12] have been found in the course of various investigations of these problems. A valuable contribution to the theory of interpolation problems was made by V. P. Potapov [13] - [18]. This article is dedicated to the results and methods of V. P. Potapov's approach.

CHAPTER I

POTAPOV'S METHOD OF SOLUTION OF INTERPOLATION PROBLEMS

1. SOME INFORMATION FROM J-ALGEBRA

1.1. The main definitions

Let us define an indefinite scalar product $[\cdot,\cdot]$ on the space of complex vectors with n components $f = col[\xi_1, \cdots, \xi_n]$ by

$$[f,g] = g^* jf,$$

where j is a symmetry matrix, i.e. a matrix with the properties

$$j^* = j; \; j^2 = I.$$

A matrix w is called j-expansive if

$$w^* j w - j \geq 0. \tag{1.1}$$

A matrix w is called j-contractive if

$$w^* j w - j \leq 0. \tag{1.2}$$

A matrix w is called j-unitary if

$$w^* j w - j = 0. \tag{1.3}$$

The expression

$$w^* j w - j = [w^* \quad I] J \begin{bmatrix} w \\ I \end{bmatrix}; \; J = \begin{bmatrix} j & 0 \\ 0 & -j \end{bmatrix}$$

is called the J-form of the matrix w.

We will be investigating analytic matrix-valued functions of a scalar complex argument with values in one of the following three standard spaces:

1. right half-plane λ : $Re \; \lambda > 0$;

2. upper half-plane z : $Im \; z > 0$;

3. unit disc s : $|s| < 1$.

The choice of standard space is not significant as one can map from one space to another. Consider, for example, the upper half-plane. The matrix-valued functions $w(z)$ which are holomorphic in the upper half-plane and satisfy the inequality

$$\frac{w(z) - w^*(z)}{i} \geq 0 \;\; (Im \; z > 0) \tag{1.4}$$

are called Nevanlinna matrix functions. We denote by $\mathcal{R}$ the class of Nevanlinna functions w extended to the lower half plane via the symmetry relation

$$w(z) \equiv w^*(\bar{z}) \tag{1.5}$$

The matrix-valued functions $w(z)$ which are holomorphic in the upper half plane and satisfy the inequality

$$w(z) + w^*(z) \geq 0 \quad (Im\ z > 0) \tag{1.6}$$

are said to be positive. For positive matrix-valued functions the symmetry relation

$$-w(z) \equiv w^*(\bar{z}) \quad (Im\ z \neq 0) \tag{1.7}$$

holds. By a generalized matrix function we mean a pair of square matrix functions of order m meromorphic in the region under consideration.

A generalized analytic matrix function is an analogue of homogeneous coordinates, which allows one to handle improper elements and regular matrix functions in one common framework.

The pair $\begin{bmatrix} P(z) \\ Q(z) \end{bmatrix}$ of matrix functions meromorphic in the upper half plane is called nondegenerate if it satisfies the nondegeneracy condition

$$\det\ [P^*(z)\quad Q^*(z)]\begin{bmatrix} P(z) \\ Q(z) \end{bmatrix} \neq 0. \tag{1.8}$$

Nondegenerate pairs $\begin{bmatrix} P(z) \\ Q(z) \end{bmatrix}$ and $\begin{bmatrix} p(z) \\ q(z) \end{bmatrix}$ are called equivalent if there is an invertible meromorphic matrix-function $T(z)$ in $Im\ z > 0$, such that

$$\begin{bmatrix} p(z) \\ q(z) \end{bmatrix} = \begin{bmatrix} P(z) \\ Q(z) \end{bmatrix} T(z).$$

A nondegenerate pair $\begin{bmatrix} P(z) \\ Q(z) \end{bmatrix}$ of matrix functions meromorphic in $Im\ z > 0$ is called J-expansive if

$$[P^*(z)\quad Q^*(z)]\,J\begin{bmatrix} P(z) \\ Q(z) \end{bmatrix} \geq 0,\ J = \begin{bmatrix} -j & 0 \\ 0 & j \end{bmatrix}.$$

If $J = \begin{bmatrix} 0 & iI \\ -iI & 0 \end{bmatrix}$ then a pair $\begin{bmatrix} P(z) \\ Q(z) \end{bmatrix}$ is called a Nevanlinna pair.

1.2. The inequality of Schwarz-Pick (S-P)

The inequality of S-P has a significant role in the solution of interpolation problems. The essence of the Potapov's method is to transfer S-P inequality via appropriate transformations to the so-called Fundamental Matrix Inequality (F.M.I.) which encodes all the information of a given problem. The S-P inequality can be considered as a generalization of Schwarz's lemma from systems of 2 points to systems of n points. Let us now deduce the S-P inequality.

THEOREM 1.1. *If* $w(z) \in \mathcal{R}$ *and* $z_1, z_2, \cdots, z_n$ *are points in the complex plane such that* $z_i \neq \bar{z}_k$ $(1 \leq i, k \leq n)$ *then the inequality*

$$\begin{bmatrix} \dfrac{w(z_1) - w^*(z_1)}{z_1 - \bar{z}_1} & \cdots & \dfrac{w(z_1) - w^*(z_n)}{z_1 - \bar{z}_n} \\ \vdots & & \vdots \\ \dfrac{w(z_n) - w^*(z_1)}{z_n - \bar{z}_1} & \cdots & \dfrac{w(z_n) - w^*(z_n)}{z_n - \bar{z}_n} \end{bmatrix} \geq 0 \qquad \text{(S-P)}$$

holds.

PROOF: We use the integral representation of a Nevanlinna matrix function

$$w(z) = \alpha + \beta z + \int_{-\infty}^{+\infty} \left(\frac{1}{t - z} - \frac{t}{1 + t^2} \right) d\sigma(t) \quad (Im\ z \neq 0)$$

where

$$\alpha = \alpha^*; \ \beta \geq 0; \ d\sigma(t) \geq 0; \ \int_{-\infty}^{+\infty} \frac{d\sigma(t)}{1 + t^2} < \infty.$$

Let us calculate the value of the expression

$$\frac{w(z_i) - w^*(z_k)}{z_i - \bar{z}_k} = \beta + \int_{-\infty}^{+\infty} \frac{d\sigma(t)}{(t - z_i)(t - \bar{z}_k)}.$$

If $f_1, f_2, \cdots, f_n$ are arbitrary vectors in the space $\mathbb{C}^m$ (column vectors with m complex entries) then we have the inequality

$$\sum_{i,k=1}^{n} f_i^* \frac{w(z_i) - w^*(z_k)}{z_i - \bar{z}_k} f_k =$$

$$= (\sum_{i=1}^{n} f_i^*) \beta (\sum_{i=1}^{n} f_i) + \int_{-\infty}^{+\infty} \left(\sum_{i=1}^{n} \frac{f_i^*}{t - z_i} \right) d\sigma(t) \left(\sum_{i=1}^{n} \frac{f_i}{t - z_i} \right) \geq 0.$$

The S-P inequality is an immediate consequence of this inequality.

1.3. Lemma on nonnegative block matrices

We shall see that the interpolation problem can be reduced to solving Fundamental Matrix Inequalities. For a description of the set of solutions of a Fundamental Matrix Inequality, we need the following lemma.

LEMMA. *(on nonnegative block matrices) The inequality*

$$\begin{bmatrix} A & B \\ B^* & C \end{bmatrix} \geq 0$$

is equivalent to the following set of conditions:

1) The inequality

$$A \geq 0$$

  holds.

2) The matrix equation

$$AX = B$$

  is solvable.

3) Each solution X of the equation in (2) satisfies

$$C - X^* AX \geq 0.$$

COROLLARY. *When A^{-1} exists, then the inequality*

$$\begin{bmatrix} A & B \\ B^* & C \end{bmatrix} \geq 0$$

is equivalent to the collection of two inequalities

1) $A > 0$;

2) $C - B^* A^{-1} B \geq 0$.

2. NEVANLINNA-PICK PROBLEM

2.1. FORMULATION OF THE PROBLEM

Among the many possible interpolation problems one might consider we begin with the Nevanlinna-Pick problem. The Nevanlinna-Pick interpolation problem is the simplest representative of the so-called classical interpolation problems of analysis. It can be formulated as follows. We are given a sequence of points

$$z_1, z_2, \cdots, z_r, \cdots \quad (Im\ z_j > 0,\ j = 1, 2, \cdots)$$

from the open upper half-plane and a collection of square matrices of size $m \times m$

$$w_1, w_2, \cdots, w_r, \cdots \quad \left(\frac{w_j - w_j^*}{i} > 0,\ j = 1, 2, \ldots\right).$$

The goal is to describe the set of Nevanlinna matrix functions satisfying the interpolation conditions

$$w(z_j) = w_j \quad (j = 1, 2, \cdots). \tag{2.1}$$

The points $z_1, z_2, \cdots, z_n, \cdots$ are called the interpolation nodes while the matrices $w_1, w_2,$ $\cdots, w_n, \cdots$ are called the interpolation values. The interpolation nodes and interpolation values together are called the interpolation data.

2.2. THE FUNDAMENTAL MATRIX INEQUALITY FOR THE NEVANLINNA-PICK PROBLEM

Let us derive the Fundatmental Matrix Inequality for the problem for a finite number of nodes.

THEOREM 2.1 [4]. *Let $w(z)$ belong to the class $\mathcal{R}$. Then in order that the matrix function $w(z)$ be a solution of the Nevanlinna-Pick problem associated with the data set as given above, it is necessary and sufficient that $w(z)$ satisfy the Fundamental Matrix Inequality*

$$\begin{bmatrix} \left[\dfrac{w_i - w_k^*}{z_i - \bar{z}_k}\right]_{i,k=1}^n & \left[\dfrac{w(z) - w_i}{z - z_i}\right]_{i=1,1}^n \\ \left[\dfrac{w^*(z) - w_k^*}{\bar{z} - \bar{z}_k}\right]_{1,k=1}^n & \dfrac{w(z) - w^*(z)}{z - \bar{z}} \end{bmatrix} \geq 0. \tag{2.2}$$

PROOF: *Necessity.* Suppose that a matrix function $w(z)$ of class $(\mathcal{R})$ satisfies the interpolation conditions

$$w(z_j) = w_j \quad (Im\ z_j \geq 0, 1 \leq j \leq n).$$

Write the Schwarz-Pick inequality (S-P) for the points $z_1, z_2, \cdots, z_n$ equal to interpolation nodes and the point $\bar{z}$. Then taking into account the conditions (2.1) and the equality $w^*(\bar{z}) = w(z)$, one gets (2.2).

Sufficiency. Let the matrix function $w(z)$ satisfy (2.2). Then nonnegativity of the submatrices in (2.2) implies the following:

1. The matrix function $w(z)$ belongs to the Nevanlinna class in $Im\ z > 0$. Indeed this fact follows from the inequality

$$\frac{w(z) - w^*(z)}{z - \bar{z}} \geq 0;$$

2. The matrix function $w(z)$ has the given value w_i at the points z_i $(i = 1, 2, \cdots)$, that is

$$w(z_i) = w_i \quad (i = 1, 2, \cdots, n)$$

Indeed we can deduce these equalities from the nonnegativity of submatrix

$$\begin{bmatrix} \dfrac{w_i - w_i^*}{z_i - \bar{z}_i} & \dfrac{w(z) - w_i}{z - z_i} \\[2mm] \dfrac{w^*(z) - w_i^*}{\bar{z} - \bar{z}_i} & \dfrac{w(z) - w^*(z)}{\bar{z} - z} \end{bmatrix} \geq 0 \ (i = 1, 2, \cdots, n).$$

From Theorem 2.1 we see that nonnegativity of the so-called information block

$$S = \left[\frac{w_i - w_k^*}{z_i - \bar{z}_k} \right]_{i,k=1}^{n}$$

is a necessary condition for solvability of the Nevanlinna-Pick problem.

2.3. SOLUTION OF THE FUNDAMENTAL MATRIX INEQUALITY

If the information block is nondegenerate (i.e. $S > 0$), then such a problem is called nondegenerate. If $det\ S = 0$, then the problem is called degenerate. The nondegenerate Nevanlinna-Pick problem has infinitely many solutions; a complete description of the set of all solutions can be given as follows.

THEOREM 2.2. *The collection of all solutions* $\{w(z)\}$ *of the Nevanlinna-Pick problem admits a parametrized description by means of a fractional-linear transformation*

$$w(z) = [a(z)\mathcal{P}(z) + b(z)Q(z)][c(z)\mathcal{P}(z) + d(z)Q(z)]^{-1} \tag{2.3}$$

where the free parameter $\begin{bmatrix} \mathcal{P}(z) \\ Q(z) \end{bmatrix}$ *runs over the set of nondegenerate, Nevanlinna pairs; the matrix* $\mathfrak{A}(z) = \begin{bmatrix} a(z) & b(z) \\ c(z) & d(z) \end{bmatrix}$ *of the fractional-linear transformation (2.3) has the form*

$$\mathfrak{A}(z) = I + i \begin{bmatrix} \dfrac{\begin{bmatrix} w_1^* \\ I \end{bmatrix}}{z - \bar{z}_n} & \cdots & \dfrac{\begin{bmatrix} w_n^* \\ I \end{bmatrix}}{z - \bar{z}_n} \end{bmatrix} S^{-1} \begin{bmatrix} w_1 & I \\ \vdots & \vdots \\ w_n & I \end{bmatrix} J; \ J = \begin{bmatrix} 0 & iI \\ -iI & 0 \end{bmatrix}. \tag{2.4}$$

PROOF: *Necessity.* Suppose that a function $w(z)$, analytic in $Im\ z > 0$, satisfies (FMI). If $S > 0$, then FMI is equivalent, by the Lemma on block-matrices, to the inequality

$$[w^*(z) \quad I] J \begin{bmatrix} w(z) \\ I \end{bmatrix}$$

$$- [w^*(z) \quad I] \left\{ -i(z - \bar{z})J \begin{bmatrix} \dfrac{\begin{bmatrix} w_1^* \\ I \end{bmatrix}}{z - \bar{z}_1} & \cdots & \dfrac{\begin{bmatrix} w_n^* \\ I \end{bmatrix}}{z - \bar{z}_n} \end{bmatrix} S^{-1} \begin{bmatrix} \dfrac{[w_1 \quad I]}{z - z_1} \\ \vdots \\ \dfrac{[w_n \quad I]}{z - z_n} \end{bmatrix} J \right\} \begin{bmatrix} w(z) \\ I \end{bmatrix} \geq 0. \tag{2.5}$$

Let us now introduce the matrix function

$$\mathfrak{A}(z) = I + i \begin{bmatrix} \dfrac{\begin{bmatrix} w_1^* \\ I \end{bmatrix}}{z - \bar{z}_1} & \cdots & \dfrac{\begin{bmatrix} w_n^* \\ I \end{bmatrix}}{z - \bar{z}_n} \end{bmatrix} S^{-1} \begin{bmatrix} w_1 & I \\ \vdots & \vdots \\ w_n & I \end{bmatrix} J.$$

The matrix $\mathfrak{A}(z)$ is J-expansive in the upper half-plane $Im\ z > 0$ and J-unitary on the boundary. It is not difficult to check that the J-form of the matrix-function $\mathfrak{A}^{-1}(z) = J\mathfrak{A}^*(\bar{z})J$

coincides with the expression included in the curly brackets. Therefore the inequality (2.5) can be written as

$$[w^*(z) \quad I]\,\mathfrak{A}^{*^{-1}}(z)J\mathfrak{A}^{-1}(z)\begin{bmatrix} w(z) \\ I \end{bmatrix} \geq 0 \qquad (2.6)$$

Let us define the pair of matrix functions $\begin{bmatrix} \mathcal{P}(z) \\ Q(z) \end{bmatrix}$ by the equality

$$\begin{bmatrix} \mathcal{P}(z) \\ Q(z) \end{bmatrix} = \mathfrak{A}^{-1}(z)\begin{bmatrix} w(z) \\ I \end{bmatrix}. \qquad (2.7)$$

The matrix functions $\mathcal{P}(z), Q(z)$ are meromorphic in $Im\ z > 0$. From (2.6) it follows that the condition

$$[\mathcal{P}^*(z) \quad Q^*(z)]\,J\begin{bmatrix} \mathcal{P}(z) \\ Q(z) \end{bmatrix} \geq 0 \quad (Im\ z > 0)$$

is satisfied. In addition the strict inequality

$$[\mathcal{P}^*(z) \quad Q^*(z)]\begin{bmatrix} \mathcal{P}(z) \\ Q(z) \end{bmatrix} = [w^*(z) \quad I]\,\mathfrak{A}^{*^{-1}}(z)\mathfrak{A}^{-1}(z)\begin{bmatrix} w(z) \\ I \end{bmatrix} > 0$$

holds for all points z, where $w(z)$ and $\mathfrak{A}^{-1}(z)$ exist. It follows from (2.7) that

$$\begin{bmatrix} w(z) \\ I \end{bmatrix} = \mathfrak{A}(z)\begin{bmatrix} \mathcal{P}(z) \\ Q(z) \end{bmatrix}. \qquad (2.8)$$

Dividing $\mathfrak{A}(z)$ into blocks

$$\mathfrak{A}(z) = \begin{bmatrix} a(z) & b(z) \\ c(z) & d(z) \end{bmatrix}$$

we obtain

$$w(z) = [a(z)\mathcal{P}(z) + B(z)Q(z)][c(z)\mathcal{P}(z) + d(z)Q(z)]^{-1} \qquad (2.9)$$

where $\begin{bmatrix} \mathcal{P}(z) \\ Q(z) \end{bmatrix}$ is a nondegenerate Nevanlinna pair in $Im\ z > 0$

Sufficiency. Suppose, conversely, that $\begin{bmatrix} \mathcal{P}(z) \\ Q(z) \end{bmatrix}$ is an arbitrary nondegenerate, Nevanlinna pair of matrix functions. Put

$$\begin{bmatrix} u(z) \\ v(z) \end{bmatrix} = \mathfrak{A}(z)\begin{bmatrix} \mathcal{P}(z) \\ Q(z) \end{bmatrix}. \qquad (2.10)$$

If $\mathfrak{A}(z)$ is obtained according to formula (2.6) and $\begin{bmatrix} P(z) \\ Q(z) \end{bmatrix}$ is a Nevanlinna pair, then the condition

$$\det v(z) \not\equiv 0 \tag{2.11}$$

is satisfied. Suppose that condition (2.11) is not satisfied, that is, $\det v(z) \equiv 0$. Then for each point z_0 in the upper half-plane there exists a vector $\vec{e} = \vec{e}(\bar{z}_0) \neq 0$ such that $v(\bar{z}_0)\vec{e}^* = 0$. It follows from (2.10) that

$$
\vec{e}\,[P^*(z_0) \quad Q^*(z_0)]\,J \begin{bmatrix} P(z_0) \\ Q(z_0) \end{bmatrix}\vec{e}^* =
$$

$$
= \vec{e}\,[u^*(z_0) \quad v^*(z_0)]\,\mathfrak{A}^{*-1}(z_0)J\mathfrak{A}^{-1}(z_0)\begin{bmatrix} u(z_0) \\ v(z_0) \end{bmatrix}\vec{e}^* =
$$

$$
= -[\vec{e}\,u^*(z_0) \quad 0]\,\{J - \mathfrak{A}^{*-1}(z_0)J\mathfrak{A}^{-1}(z_0)\}\begin{bmatrix} u(z_0)\vec{e}^* \\ 0 \end{bmatrix} + [\vec{e}\,u^*(z_0) \quad 0]\,J\begin{bmatrix} u(z_0)\vec{e}^* \\ 0 \end{bmatrix} =
$$

$$
= -[\vec{e}\,u^*(z_0) \quad 0]\,\{J - \mathfrak{A}^{*-1}(z_0)J\mathfrak{A}^{-1}(z_0)\}\begin{bmatrix} u(z_0)\vec{e}^* \\ 0 \end{bmatrix} \geq 0.
$$

Taking into consideration that z_0 is any point in the upper half-plane and that $J - \mathfrak{A}^{*-1}(z)J\mathfrak{A}^{-1}(z) > 0$, we have

$$u(z_0)\vec{e}^* = 0.$$

Therefore it is necessary that

$$
\vec{e}\,[P^*(z_0) \quad Q^*(z_0)]\begin{bmatrix} P(z_0) \\ Q(z_0) \end{bmatrix}\vec{e}^*
$$

$$
= \vec{e}\,[u^*(z_0) \quad v^*(z_0)]\,\mathfrak{A}^{*-1}(z_0)\mathfrak{A}^{-1}(z_0)\begin{bmatrix} u(z_0) \\ v(z_0) \end{bmatrix}\vec{e}^* = 0.
$$

Since $\begin{bmatrix} P(z) \\ Q(z) \end{bmatrix}$ is a nondegenerate pair, we have

$$\det v(z) \not\equiv 0.$$

Furthermore, we have

$$w(z) = u(z)v^{-1}(z) =$$

$$= [a(z)P(z) + b(z)Q(z)][c(z)P(z) + d(z)Q(z)]^{-1}.$$

Let us show that the matrix function $w(z)$ is a solution of FMI. If $\begin{bmatrix} P(z) \\ Q(z) \end{bmatrix}$ is a Nevanlinna pair, then the inequality

$$[\mathcal{P}^*(z) \quad Q^*(z)] \, J \begin{bmatrix} P(z) \\ Q(z) \end{bmatrix} =$$

$$= v^{-1*}(z) [w^*(z) \quad I] \, \mathfrak{A}^{*-1}(z) J \mathfrak{A}^{-1}(z) \begin{bmatrix} w(z) \\ I \end{bmatrix} v^{-1}(z) \geq 0 \qquad (2.12)$$

is satisfied. We remark that the left side of (2.12) is equivalent to inequality (2.6). Thus, the fractional-linear transformation (2.3) maps the set of all Nevanlinna pairs $\begin{bmatrix} P(z) \\ Q(z) \end{bmatrix}$ into the set $\{w(z)\}$ of all solutions of inequality (FMI).

From Theorem 2.2 it follows that $S > 0$ is a sufficient condition for the solvability of the Nevanlinna-Pick problem. In Chapter II, we shall establish that the inequality $S \geq 0$ is a necessary and sufficient condition for the solvability of the Nevanlinna-Pick problem.

3. SCHUR PROBLEM

3.1. The formulation of problem

We are given square matrices $c_0, c_1, \cdots, c_{n-1}$ of size $m \times m$ and wish to describe the set of matrix functions $w(s)$, holomorphic in the unit disc, which satisfy the inequality

$$I - w^*(s)w(s) \geq 0 \text{ for } |s| < 1 \qquad (3.1)$$

together with the interpolation condition

$$w(s) = c_0 + c_1 s + \cdots + c_{n-1} s^{n-1} + \cdots, \qquad (3.2)$$

i.e. the first $n + 1$ coefficients of the Maclaurin series for $w(s)$ are given matrices.

3.2. The Fundamental Matrix Inequality

THEOREM 3.1. *Let $w(s)$ be holomorphic in the unit disc. Then in order that the matrix function $w(s)$ be a solution of the Schur problem it is necessary and sufficient that $w(s)$ satisfy the Fundamental Matrix Inequality*

$$
\left[
\begin{array}{ccccc}
I - c_0 c_0^* & -c_0 c_1^* & \cdots & -c_0 c_{n-1}^* & \dfrac{w(s) - c_0}{s} \\[2mm]
-c_1 c_0^* & I - c_0 c_0^* - c_1 c_1^* & \cdots & -c_1 c_{n-1}^* - c_0 c_{n-2}^* & \dfrac{w(s) - (c_0 + c_1 s)}{s^2} \\[2mm]
\vdots & \vdots & \vdots & \vdots & \vdots \\[2mm]
-c_{n-1} c_0^* & -c_{n-1} c_1^* - c_{n-2} c_0^* & \cdots & I - c_0 c_0^* - c_1 c_1^* \cdots - c_{n-1} c_{n-1}^* & \dfrac{w(s) - \sum_{i=0}^{n-1} c_i s^i}{s^n} \\[2mm]
\dfrac{w^*(s) - c_0^*}{\bar{s}} & \dfrac{w^*(s) - (c_0^* + c_1^* \bar{s})}{\bar{s}^2} & \cdots & \dfrac{w^*(s) - \sum_{i=0}^{n-1} c_i^* s^i}{\bar{s}^n} & \dfrac{I - w^*(s)w(s)}{1 - \bar{s}s}
\end{array}
\right] \geq 0.
$$

$$\tag{3.3}$$

PROOF: *Necessity.* Suppose that a matrix-function $w(s)$ is a solution of the Schur problem. We write the Schwarz-Pick inequality for the system of $(n+1)$ points $s_1, s_2, \cdots, s_n$, from the unit disc and a point $\frac{1}{\bar{s}}$. We then multiply this inequality on the right by the matrix

$$
T =
\left[
\begin{array}{ccccc}
\dfrac{I}{\varphi_1'(s_1)} & \dfrac{I}{\varphi_2'(s_1)} & \cdots & \dfrac{I}{\varphi_n'(s_n)} & 0 \\[2mm]
0 & \dfrac{I}{\varphi_2'(s_2)} & \cdots & \dfrac{I}{\varphi_n'(s_n)} & 0 \\[2mm]
\vdots & \vdots & & \vdots & \vdots \\[2mm]
0 & 0 & \cdots & \dfrac{I}{\varphi_n'(s_n)} & 0 \\[2mm]
0 & 0 & \cdots & 0 & I
\end{array}
\right]
$$

where

$$
\varphi_i(s) = (s - s_1)(s - s_2) \cdots (s - s_i) \quad (i = 1, 2, \cdots, n)
$$

and on the left by the matrix T^*, and then take a limit as $s_i \to 0 \ (i = 1, 2, \cdots, n)$. Taking into consideration the interpolation condition (3.2) and the symmetry relation $w^* \left(\dfrac{1}{\bar{s}} \right) = w(s)$, we recover the Fundatmental Matrix Inequality of the Schur problem (3.3).

Sufficiency. Suppose that the matrix function $w(s)$ satisfies the Fundamental Matrix Inequality of the Schur problem (3.3). Firstly the nonnegativity of the matrix in (3.3) implies that

$$
\dfrac{I - w^*(s)w(s)}{1 - \bar{s}s} \geq 0 \quad (|s| < 1)
$$

that is, $w(s)$ is contractive in unit disc. Secondly, for each κ the submatrix

$$
\begin{bmatrix}
I - \sum\limits_{i=1}^{\kappa} c_i c_i^* & \dfrac{w(s) - \sum\limits_{i=0}^{\kappa} c_i s^i}{s^{\kappa+1}} \\
* & \dfrac{I - w^*(s)w(s)}{1 - \bar{s}s}
\end{bmatrix} \geq 0 \quad (\kappa = 0, 1, \cdots, n-1).
$$

is nonnegative. Therefore $w(s)$ satisfies the interpolation condition (3.2). The assertion of the Theorem now follows.

3.3. Solution of Fundamental Matrix Inequality

THEOREM 3.2. *If the information block of the Schur problem is nondegenerate, then the Schur problem has infinitely many solutions. Every generalized solution of the FMI for the Schur problem has the fractional-linear form*

$$
w(s) = [a(s)\mathcal{P}(s) + b(s)\mathcal{Q}(s)][c(s)\mathcal{P}(s) + d(s)\mathcal{Q}(s)]^{-1}
$$

where the coefficient matrix $\mathfrak{A}(s)$ is given by a formula

$$
\mathfrak{A}(s) = \begin{bmatrix} a(s) & b(s) \\ c(s) & d(s) \end{bmatrix} = I + (1-s)[\,I\ I\ \ \cdots\ \ I\,]
$$

$$
\times \begin{bmatrix}
I & 0 & \cdots & 0 \\
-c_0 & -c_1 & \cdots & -c_{n-1} \\
0 & I & \cdots & 0 \\
0 & -c_0 & \cdots & -c_{n-2} \\
\vdots & \vdots & & \vdots \\
0 & 0 & \cdots & I \\
0 & 0 & \cdots & -c_0
\end{bmatrix} S^{-1}
\begin{bmatrix}
I & -c_0^* & 0 & 0 & \cdots & 0 & 0 \\
0 & -c_1^* & I & -c_0^* & \cdots & 0 & 0 \\
\vdots & \vdots & \vdots & \vdots & & \vdots & \vdots \\
0 & -c_{n-1}^* & 0 & -c_{n-2}^* & \cdots & I & -c_0^*
\end{bmatrix}
\begin{bmatrix}
I \\
sI \\
\vdots \\
s^{n-1}I
\end{bmatrix} J
$$

where

$$
J = \begin{bmatrix} -j & 0 \\ 0 & j \end{bmatrix}.
$$

The free paramenter $\begin{bmatrix} \mathcal{P}(s) \\ \mathcal{Q}(s) \end{bmatrix}$ runs over the set of all nondegenerate, J-contractive pairs which are meromorphic in $|s| \neq 1$.

A proof of this theorem is only a little different from the one in 2.3.

A solvability condition for the Schur problem can be formulated in terms of interpolation data. More precisely, the nonnegativity of the so-called information block

$$S = \begin{bmatrix} I - c_0 c_0^* & -c_0 c_1^* & \cdots & -c_0 c_{n-1}^* \\ -c_1 c_0^* & I - c_0 c_0^* - c_1 c_1^* & \cdots & -c_1 c_{n-1}^* - c_0 c_{n-2}^* \\ \vdots & \vdots & & \vdots \\ -c_{n-1} c_0^* & -c_{n-1} c_1^* - c_{n-2} c_0^* & \cdots & I - c_0 c_0^* - c_1 c_1^* - \cdots - c_{n-1} c_{n-1}^* \end{bmatrix} \geq 0$$

is a necessary and sufficient condition for solvability of the Schur problem.

CHAPTER II

OPERATOR IDENTITIES AND INTERPOLATION PROBLEMS

INTRODUCTION

While investigating a whole class of interpolation problems, Potapov and his associates solved each new type of interpolation problem separately by identifying common traits with previously solved versions; no description of the set of all solutions or efficient procedure for reduction to the corresponding FMI was found for a general class. Each type of interpolation was investigated separately.

In this chapter we introduce a method of operator identities which serves to unify the particular instances of Potapov's method under one framework. Following this method, we identify a general interpolation problem containing all the classical interpolation problems as special cases, and write down the Potapov inequality and the description of the set of all solutions in this general setting.

1. FORMULATION OF THE PROBLEM

Introduce the Hilbert spaces H, G and G_1, where

$$G = G_1 \oplus G_1 \quad (dim\ G_1 < \infty) \tag{1.1}$$

Suppose that the operators A, Π, J, S are connected by the operator identity

$$AS - SA^* = i\Pi J \Pi^* \tag{1.2}$$

where

$$A, S \in \{H; H\}; \Pi \in \{G; H\}; \; J \in \{G; G\}; J^2 = E; \; J^* = J.$$

The symbol $\{H_1; H_2\}$ designates the set of bounded operators acting from H_1 to H_2. The operator Π has the representation $\Pi = [\Pi_1 \quad \Pi_2]$, where $\Pi_1, \Pi_2 \in \{G_1; H\}$.

Next we assume that the block representation of the operator J corresponding to the decomposition (1.1) has the form

$$J = \begin{bmatrix} 0 & E_1 \\ E_1 & 0 \end{bmatrix}$$

where E_1 is the identity operator on G_1.

Let us denote by $\mathcal{N}$ the operators in the class $\{H; H\}$ which have a finite or countably many points of spectrum. Suppose that $A \in \mathcal{N}$.

Let $\mathcal{E}$ be a set of monotonically increasing operator functions $\sigma(t) \in \{G_1; G_1\}$ such that the integrals

$$S_\sigma = \int_{-\infty}^{+\infty} (E - At)^{-1} \Pi_2 [d\sigma(t)] \Pi_2^* (E - A^* t)^{-1} \tag{1.4}$$

and

$$J = \int_{-\infty}^{+\infty} \frac{d\sigma(t)}{1 + t^2} \tag{1.5}$$

converge in a weak sense. Then the integral

$$\Pi_{1,\sigma} = -i \int_{-\infty}^{\infty} [A(E - tA)^{-1} + \frac{t}{1 + t^2} E] \Pi_2 d\sigma(t) \tag{1.6}$$

also converges in a weak sense.

Let us introduce the operators

$$\tilde{S} = S_\sigma + FF^*, \quad \tilde{\Pi}_1 = \Pi_{1,\sigma} + i[\Pi_2 \alpha + F\beta^{\frac{1}{2}}] \tag{1.7}$$

where

$$\alpha, \beta \in \{G_1; G_1\}, \quad \alpha = \alpha^*, \quad \beta \geq 0$$

$$AF = \pi_2 \beta^{\frac{1}{2}}, \; F \in \{G_1, H\}. \tag{1.8}$$

The operator identity (1.21) suggests the following interpolation problem:

Describe the set of $\sigma(t) \in \mathcal{E}$ and $\alpha = \alpha^*$, $\beta \geq 0$, which gives the representations

$$S = \tilde{S}, \quad \Pi_1 = \tilde{\Pi}_1 \tag{1.9}$$

where S and Π_1 are given operators and $\tilde{S}, \tilde{\Pi}$ are as in (1.7). A necessary condition for the solvability of this problem is the inequality

$$S \geq 0. \tag{1.10}$$

Later we shall prove that the inequality $S \geq 0$ is a sufficient condition as well.

LEMMA 1.1. *If $A \in \mathcal{N}$ and $\sigma(t) \in \mathcal{E}$, then the operator $\tilde{S}$ satisfies an operator identity*

$$A\tilde{S} - \tilde{S}A^* = i\tilde{\Pi}J\tilde{\Pi}^*; \quad \tilde{\Pi} = \begin{bmatrix} \tilde{\Pi}_1 & \Pi_2 \end{bmatrix} \tag{1.11}$$

We deduce the following result from Lemma 1.1 by linear algebra methods.

LEMMA 1.2. *If $\Pi_2 g \neq 0$ whenever $g \neq 0$ and the representation $S = \tilde{S}$ holds, then the representation $\Pi_1 = \tilde{\Pi}_1$ also holds.*

Let us denote by $\mathcal{K}$ the class of operators A, for which the equation $AS - SA^* = 0$ has only the trivial solution $S = 0$ in $\{H; H\}$.

LEMMA 1.3. *If $A \in \mathcal{K}$ and $\Pi_1 = \tilde{\Pi}_1$, then $S = \tilde{S}$.*

2. THE FUNDAMENTAL MATRIX INEQUALITY

Let $\mathcal{D}$ be the set of complex numbers such that $z \neq \bar{z}$ and z^{-1} does not belong to the spectrum of the operator A. The operator identity (1.2) corresponds to the matrix inequality

$$L(z) = \begin{bmatrix} S & B(z) \\ B^*(z) & C(z) \end{bmatrix} \geq 0, \ z \in \mathcal{D} \tag{2.1}$$

where

$$B(z) = (E - Az)^{-1}(\Pi_1 - i\Pi_2 w(z)); \ C(z) = \frac{w(z) - w^*(z)}{z - \bar{z}}. \tag{2.2}$$

The solution of inequality (2.1) should be found in the class $\mathcal{R}$ of analytic operator functions $w(z)$ such that

$$w^*(\bar{z}) = w(z); \quad (w(z) - w^*(z))/(z - \bar{z}) \geq 0. \tag{2.3}$$

for all $z \neq \bar{z}$. The inequality (2.1) is the abstract analogue of the Potapov Fundamental Matrix Inequalities.

An operator function $w \in \mathcal{R}$ has the representation

$$w(z) = \beta z + \alpha + \int_{-\infty}^{+\infty} \left(\frac{1}{t-z} - \frac{t}{1+t^2} \right) d\sigma(t) \tag{2.4}$$

where $\alpha, \beta \in \{G_1; G_1\}$, $\alpha = \alpha^*$, $\beta \geq 0$ and $\sigma(t) \in \{G_1; G_1\}$ is a monotonically increasing operator function.

The formula (2.4) for the operator function $w(z)$ suggests the solution of the abstract problem formulated in §1, and leads to the investigation of the following interpolation problem.

We are given operators A, Π, J, S, connected by operator identity (1.2) and wish to describe the collection of $w(z) \in \mathcal{R}$ such that $\sigma(t), \alpha, \beta$ are from representation (2.4) and satisfy the identities (1.9), and the operators $\Pi_1, \tilde{S}$ are determined by the equalities (1.4), (1.5) and (1.9).

THEOREM 2.1. *Let $A \in \mathcal{N}$ and the operator function $w(z)$ be determined by the formula (2.4). If $\sigma(t) \in \mathcal{E}$ and the operator $F \in \{G_1; H\}$ are such that S given by (1.9) and (1.7) satisfies (1.10), then the inequality*

$$\tilde{L}(z) = \begin{bmatrix} S & \tilde{B}(z) \\ \tilde{B}^*(z) & C(z) \end{bmatrix} \geq 0 \tag{2.5}$$

holds, where

$$\tilde{B}(z) = (E - Az)^{-1}(\tilde{\Pi}_1 - i\Pi_2 w(z)). \tag{2.6}$$

PROOF: One can check the identity

$$\left(\begin{bmatrix} \tilde{S} & \tilde{B}(z) \\ \tilde{B}^*(z) & C(z) \end{bmatrix} \begin{bmatrix} f \\ g \end{bmatrix}, \begin{bmatrix} f \\ g \end{bmatrix} \right) = [f + g, \quad f + g]_z \tag{2.7}$$

by a simple calculation. In this identity the symbol $[\cdot, \cdot]_z$ denotes the scalar product of the vectors from $H + G_1$ determined by the equations

$$[f, f]_z = \int_{-\infty}^{+\infty} ([d\sigma(t)]\Pi_2^*(E - A^*t)^{-1}f, \ \Pi_2^*(E - A^*t)^{-1}f) + (F^*f, F^*f) \tag{2.8}$$

$$[g, g]_z = \int_{-\infty}^{+\infty} \frac{(d\sigma(t)g, g)}{|t - z|^2} + (\beta^{\frac{1}{2}}g, \beta^{\frac{1}{2}}g) \tag{2.9}$$

$$[f, g]_z = \int_{-\infty}^{+\infty} \left([d\sigma(t)]\Pi_2^*(E - A^*t)^{-1}f, \ \frac{g}{t - z} \right) + (F^*f, \beta^{\frac{1}{2}}g). \tag{2.10}$$

From the formulas (2.5)-(2.10) the assertion of the theorem follows.

We say that the operator A belongs to the class $\mathcal{N}_0$, if one of the following assertions is satisfied:

1. $A \in \mathcal{N}$ and for some $\delta > 0, \epsilon > 0$ the spectrum is located in the angle

$$\delta \le |arg \ z| \le \pi - \delta, \ |z| \ge \epsilon;$$

2. The spectrum of A is concentrated at zero and the equality

$$\overline{lim}_{r \to \infty} \{[\ln M(r)]/z^\rho\} < \infty$$

holds, where

$$M(r) = \max_{0 \le \theta \le 2\pi} \|(E - Az)^{-1}\|, \ z = re^{i\theta}, \ \rho > 0.$$

THEOREM 2.2. Suppose $A \in \mathcal{N}_0$ and the operator function $w(z)$ satisfies inequality (2.1) while $\sigma(t) \in \mathcal{E}$. Then the equality

$$\Pi_2\beta = -iA(\Pi_1 - \Pi_{1,\sigma} - i\Pi_2\alpha) \tag{2.12}$$

holds, where $\sigma(t)$, α, β are the same as in the representation (2.4).

PROOF: Let us examine the function

$$w_\sigma(z) = w(z) - \beta z - \alpha = \int_{-\infty}^{+\infty} \left(\frac{1}{t-z} - \frac{t}{1+t^2} \right) d\sigma(t) \qquad (2.13)$$

Applying Theorem 2.1 in the case of (2.13) we get the equality

$$L_\sigma(z) = \begin{bmatrix} S_\sigma & B_\sigma(z) \\ B_\sigma^*(z) & C_\sigma(z) \end{bmatrix}, \quad z \in \mathcal{D} \qquad (2.14)$$

where the operator S_σ is determined by formula (1.4), and where $B_\sigma(z)$ and $C_\sigma(z)$ are given by

$$B_\sigma(z) = (E - Az)^{-1}[\Pi_{1,\sigma} - i\Pi_2 w_\sigma(z)];$$

$$C_\sigma(z) = [w_\sigma(z) - w_\sigma^*(z)]/(z - \bar{z}).$$

From the inequalities (2.1) and (2.14) we deduce that

$$\|B(z)\|^2 \leq \|S\|(C(z)g, g), \quad \|B_\sigma(z)g\| \leq \|S_\sigma\|(C_\sigma(z)g, g).$$

Hence for any $\sigma > 0$ and $\epsilon > 0$ we have

$$\|B(z)g\| = 0(1), \quad \|B_\sigma(z)g\| = 0(1), \quad \delta \leq |Arg\, z| \leq \pi - \delta, \quad |z| \geq \epsilon.$$

So the relation

$$\|[B(z) - B_\sigma(z)]g\| = \|(E - Az)^{-1}(\Pi_1 - \Pi_{1,\sigma} - i\Pi_2\alpha - i\Pi_2\beta z)\| = 0(1) \qquad (2.15)$$

is valid. As $A \in \mathcal{N}_0$ the relation (2.12) follows from (2.15).

COROLLARY 2.1. *Let $A \in \mathcal{N}_0$, and suppose that zero is not an eigenvalue of A. If $w(z)$ satisfies inequality (2.1) and $\sigma(t) \in \mathcal{E}$, then there exists unique operator $F \in \{G; H\}$ satisfying the relation (1.10); moreover the representation $\Pi_1 = \tilde{\Pi}_1$ is valid.*

Taking (1.10) into account we obtain

$$F = A^{-1}\Pi_2\beta^{\frac{1}{2}}.$$

Therefore, the formula (2.12) takes the form

$$\Pi_1 = \Pi_{1,\sigma} + i[\Pi_2\alpha + F\beta^{\frac{1}{2}}] = \tilde{\Pi}_1.$$

3. THE TRANSFORMED INEQUALITY

In Section 2 (Corollary 2.1) conditions are given which guarantee the validity of the representation $\Pi_1 = \tilde{\Pi}_1$. To obtain the representation $S = \tilde{S}$ we use the transformed inequality. Put

$$\tau_0(z) = iA(E - Az)^{-1}, \ \tau_\kappa(z) = (E - Az)^{-1}\Pi_\kappa \ (\kappa = 1, 2) \tag{3.1}$$

and consider the operator

$$L_T(z) = \begin{bmatrix} E & O \\ \tau_0(\bar{z}) & \tau_2(\bar{z}) \end{bmatrix} \begin{bmatrix} S & B(z) \\ B^*(z) & C(z) \end{bmatrix} \begin{bmatrix} E & \tau_0^*(\bar{z}) \\ 0 & \tau_2^*(\bar{z}) \end{bmatrix} \ (z \in \mathcal{D}) \tag{3.2}$$

The operator $L_T(z)$ is acting in the space $H \oplus H$ and can be written as

$$L_T(z) = \begin{bmatrix} S & B_T(z) \\ B_T^*(z) & C_T(z) \end{bmatrix} \ (z \in \mathcal{D}) \tag{3.3}$$

where

$$B_T(z) = S\tau_0^*(\bar{z}) + B(z)\tau_2^*(\bar{z}); \tag{3.4}$$

$$C_T(z) = \frac{i}{z - \bar{z}}[B_T(z) + B_T^*(z)]. \tag{3.5}$$

In view of (3.2), we have that any solution $w(z)$ of the inequality (2.1) is the solution of transformed inequality

$$L_T(z) = \begin{bmatrix} S & B_T(z) \\ B_T^*(z) & C_T(z) \end{bmatrix} \geq 0 \ (z \in \mathcal{D}) \tag{3.6}$$

We note that Potapov [18] and later Katsnel'son [6] used the transformed inequality while examining a number of concrete interpolation problems.

The inequality (3.6) is an abstract analogue of Potapov's transformed inequality. On the one hand, by (3.6), we have the estimate

$$\|B_T(z)f\|^2 \leq \|S\|(C_T(z)f, f)$$

while on the other hand, by (3.5), we have

$$(C_T(z)f, f) \le \|B_T(z)f\| \cdot \|f\|/|Im\ z|.$$

These estimates show that

$$\|B_T(z)\| \le \|S\|/|Im\ z|. \tag{3.9}$$

Finally, in view of (3.5), (3.6) and (3.9), we obtain

$$iB_T(z) = \int_{-\infty}^{+\infty} [d\sigma_T(t)]/(t - z) \tag{3.10}$$

where $\sigma_T(t)$ is a non-decreasing operator function with the values in class $\{H; H\}$ while

$$\int_{-\infty}^{+\infty} (d\sigma_T(t)f, f) \le \|S\| \cdot \|f\|^2. \tag{3.11}$$

Using the inversion formula for the integral (3.10), from (3.1) and (3.4) we deduce the equality

$$\int_{\alpha}^{\beta} d\sigma_T(t) = \int_{\alpha}^{\beta} (E - At)^{-1}\Pi_2 d\sigma(t)\Pi_2^*(E - A^*t)^{-1}. \tag{3.12}$$

In view of (3.11) and (3.12), we have

THEOREM 3.1. *If $w(z) \in \mathcal{R}$ is a solution of inequality (2.1), then the corresponding operator function $\sigma(t)$ belongs to the class $\mathcal{E}$.*

From this theorem, it follows that the integral on the right side of (1.4) is weakly convergent while $S_\sigma \le S$.

THEOREM 3.2. *If $A \in \mathcal{N}, \Pi_1 = \tilde{\Pi}_1$ and $w(z) \in \mathcal{R}$ is a solution of inequality (2.1), then*

$$A(S - \tilde{S}) = 0. \tag{3.13}$$

If zero is not an eigenvalue of A then

$$S = \tilde{S}. \tag{3.14}$$

PROOF: From (3.6), the inequality

$$\tilde{L}_T(z) = \begin{bmatrix} \tilde{S} & \tilde{B}_T(z) \\ \tilde{B}_T^*(z) & C(z) \end{bmatrix} \ge 0,\ z \in \mathcal{D} \tag{3.15}$$

holds, where

$$\tilde{B}_T(z) = \tilde{S}\tau_0^*(\bar{z}) + \tilde{B}(z)\tau_2^*(\bar{z}).$$

We remind the reader that $B(z)$ and $\tilde{B}(z)$ are defined in (2.2) amd (2.6) respectively. Since $\Pi_1 = \tilde{\Pi}_1$, then $B(z) = \tilde{B}(z)$. Taking (3.9) and an analogous estimate for $\tilde{B}_T(z)$ into account we obtain

$$\|[B_T^*(z) - \tilde{B}_T(z)]f\| = \|(E - Az)^{-1}A(S - \tilde{S})f\| = 0(1) \qquad (3.16)$$

where

$$|z| \geq \epsilon,\ 0 < \delta \leq Arg\ z < \pi - \delta.$$

From (3.16) the condition (3.13) follows. If zero is not an eigenvalue of the operator A then equality (3.14) is deduced from (3.13) and the theorem is proved.

In view of Theorems 3.1, 3.2 and Corollary 2.1, we have the main result

THEOREM 3.3. *Suppose $A \in \mathcal{N}_0$ and zero is not an eigenvalue of the operator A If the operator function $w(z) \in \mathcal{R}$ satisfies the inequality (2.4), then the corresponding operator function $\sigma(t)$ belongs to the class $\mathcal{E}$ and the integral representation (1.11) holds.*

COROLLARY 3.1. *Suppose that the conditions of Theorem 3.3 are satisfied. Then the collection $\{\sigma(t), \alpha, \beta\}$, which gives the solution of interpolation problem (1.11) coincides with the set $\{\sigma(t), \alpha, \beta\}$ which gives the solution of inequality (2.1) via formula (2.4).*

4. THE SOLUTION OF NONDEGENERATE INTERPOLATION PROBLEMS

As seen in Section 3 (Corollary 3.1) the interpolation problem (1.11) is equivalent to the Fundamental Matrix Inequality (2.1). In this paragraph we shall study the solutions of the FMI in the nondegenerate case. The interpolation problem (1.11) is said to be nondegenerate if the corresponding operator S is bounded together with its inverse. When this condition is not satisfied, then the interpolation problem (1.11) is said to be degenerate.

Suppose that the operator function $\varphi(z) \in \{G_1; G\}$ is meromorphic in the upper half-plane. Let us consider the inequality

$$\left[\begin{matrix} S & (E - Az)^{-1}\Pi J\varphi(z) \\ \varphi^*(z)J\Pi^*(E - A^*\bar{z})^{-1} & \dfrac{\varphi^*(z)J\varphi(z)}{2Im\ z} \end{matrix} \right] \geq 0. \tag{4.1}$$

We note that the inequality (4.1) coincides with inequality (2.1) if $\varphi(z) = col\ [-iw(z), E]$.

THEOREM 4.1. *Suppose that the operator S is positive and bounded together with its inverse and satisfies condition (1.2). Then in order that the operator function $\varphi(z)$ be the solution of the inequality (4.1) it is necessary and sufficient that $\varphi(z)$ admit the representation*

$$\varphi(z) = \mathfrak{A}(z)R(z) \tag{4.2}$$

where

$$\mathfrak{A}(z) = E - iz\Pi^*(E - zA^*)^{-1}S^{-1}\Pi J. \tag{4.3}$$

The free parameter $R(z)$ runs over the set of all meromorphic operator functions with values in $\{G_1; G\}$ which satisfy the inequality

$$R^*(z)JR(z) \geq 0, \quad Im\ z > 0. \tag{4.4}$$

PROOF: *Necessity.* Rewrite inequality (4.1) in the form

$$\varphi^*(z)J\varphi(z) - \varphi^*(z)\{2(Im\ z)J\Pi^*(E - A^*\bar{z})^{-1}S^{-1}(E - Az)^{-1}\Pi J\}\varphi(z) \geq 0. \tag{4.5}$$

We define the operator function $\mathfrak{A}(z)$ by the formula

$$\mathfrak{A}(z) = E - iz\Pi^*(E - zA^*)^{-1}S^{-1}\Pi J. \tag{4.6}$$

In view of (1.2), we obtain

$$\mathfrak{A}(z)J\mathfrak{A}^*(z) - J = \frac{z - \bar{z}}{i}\Pi^*(E - zA^*)^{-1}S^{-1}(E - \bar{z}A)^{-1}\Pi.$$

Taking into consideration $S > 0$, let us derive some corollaries:

1. $\mathfrak{A}(z)$ is J-expansive in $Im\ z > 0$

$$\mathfrak{A}(z)J\mathfrak{A}^*(\bar{z}) - J \geq 0;$$

2. $\mathfrak{A}(z)$ is J-unitary in $Im\ z = 0$

$$\mathfrak{A}(z)J\mathfrak{A}^*(z) - J = 0$$

where $(\bar{z})^{-1}$ does not belong to the spectrum of the operator A. From the symmetry relation and (4.6), it follows that

$$\mathfrak{A}^{-1}(z) = J\mathfrak{A}^*(\bar{z})J = E - izJ\Pi^*S^{-1}(E - zA)^{-1}\Pi$$

and the J-form of the operator funciton $\mathfrak{A}^{-1}(z)$ is obtained by the formula

$$J - \mathfrak{A}^{*-1}(z)J\mathfrak{A}^{-1}(z) = J(J - \mathfrak{A}(\bar{z})J\mathfrak{A}^*(\bar{z}))J =$$
$$= 2(Im\ z)J\Pi^*(E - A^*\bar{z})^{-1}S^{-1}(E - Az)^{-1}\Pi J. \tag{4.7}$$

Finally, in view of (4.7), we have that inequality (4.5) is equivalent to the inequality

$$\varphi^*(z)\mathfrak{A}^{*-1}(z)J\mathfrak{A}^{-1}(z)\varphi(z) \geq 0. \tag{4.8}$$

Putting

$$R(z) = \mathfrak{A}^{-1}(z)\varphi(z)$$

let us consider the operator function

$$\varphi(z) = \mathfrak{A}(z)R(z).$$

Note that the inequality

$$R^*(z)JR(z) \geq 0, \quad Im\ z > 0$$

holds.

Sufficiency. Let $R(z)$ be an arbitrary operator function meromorphic in the upper half-plane, satisfying the inequality (4.4). The operator-function $\mathfrak{A}(z)$ is given by formula (4.6). Put

$$\varphi(z) = \mathfrak{A}(z)R(z).$$

Furthermore, since

$$R(z) = \mathfrak{A}^{-1}(z)\varphi(z)$$

the inequality (4.1) results from the conditions (4.8) and (4.7). The assertion of the Theorem has been proved.

Next we give the full description of the set of solutions of the interpolation problem in the nondegenerate case.

THEOREM 4.2. *Suppose that the operator S is positive and bounded together with its inverse and satisfies the identity (1.2); assume also that $\Pi_2 g \neq 0$ if $g \neq 0$, $A \in \mathcal{N}_0$. Then in order that the operator function $w(z)$ be the solution of the inequality (2.1) it is necessary and sufficient that $w(z)$ admit the representation*

$$w(z) = i[a(z)P(z) + b(z)Q(z)][c(z)P(z) + d(z)Q(z)]^{-1} \tag{4.9}$$

where the coefficient matrix $\mathfrak{A}(z) = \begin{bmatrix} a(z) & b(z) \\ c(z) & d(z) \end{bmatrix}$ is given by the formula

$$\mathfrak{A}(z) = \begin{bmatrix} a(z) & b(z) \\ c(z) & d(z) \end{bmatrix} = E - iz\Pi^*(E - zA^*)^{-1}S^{-1}\Pi J;$$

the free parameter $\begin{bmatrix} P(z) \\ Q(z) \end{bmatrix}$ runs over the set of all nondegenerate pairs of operator functions which are meromorphic in $Im\ z > 0$ and have the J-property.

5. WEYL DISCS

Let us consider the set of Hilbert spaces H_m, where the parameter m belongs to some set of numbers $\mathcal{M}$, with the property that $H_\kappa \subset H_m$ for $\kappa < m$, $\kappa, m \in \mathcal{M}$. Denote $\overline{\cup H_m}$ by $\tilde{H}$. The parameter m might change discretely ($m = 1, 2, \cdots$) or continuously. In the first case we obtain discrete interpolation problems. In the second case we move to the so-called continuous problems. As shown in §4, solving the nondegenerate interpolation problem for $w(z)$ is equivalent to the investigation of the inequality

$$[iw^*(z) \quad E]\,\mathfrak{A}_m^{*-1}(z)J\mathfrak{A}_m^{-1}(z) \begin{bmatrix} -iw(z) \\ E \end{bmatrix} \geq 0 \tag{5.1}$$

where $w(z)$ is an analytic operator function in the upper half plane satisfying the condition (2.3).

The main question in the investigation of the interpolation problems for the limiting case $m \to \tilde{m}$ ($\tilde{m} \leq \infty$) is the question about behavior of the J-form of the operator $\mathfrak{A}_m^{-1}(z)$, as $m \to \tilde{m}$.

Let us designate the operator of orthogonal projection of $\tilde{H}$ to H_m as P_m.

THEOREM 5.1. *Let the operators* $\Pi_{1,m}, \Pi_{2,m}, A_m, S_m$ *satisfy the conditions of Theorem 4 2 and also the following identities*

$$P_\kappa S_m P_\kappa = S_\kappa; \quad P_\kappa A_m = P_\kappa A_m P_\kappa = A_\kappa; \quad P_\kappa \Pi_{1,m} = \Pi_{1,\kappa}; \quad P_\kappa \Pi_{2,m} = \Pi_{2,\kappa}.$$

Then the J-form of the operator function $\mathfrak{A}_m^{-1}(z)$ increases as m increases.

PROOF: According to the Theorem on factorization [27] we have

$$\mathfrak{A}_m^{-1}(z) = V(z)\mathfrak{A}_\kappa^{-1}(z), \quad m > \kappa$$

where $V(z)$ and $\mathfrak{A}_\kappa^{-1}(z)$ are J-contractive transformations. Consider the difference of the J-forms of $\mathfrak{A}_m^{-1}(z)$ and $\mathfrak{A}_\kappa^{-1}(z)$

$$[J - \mathfrak{A}_m^{*-1}(z)J\mathfrak{A}_m^{-1}(z)] - [J - \mathfrak{A}_\kappa^{*-1}(z)J\mathfrak{A}_\kappa^{-1}(z)] =$$

$$= \mathfrak{A}_\kappa^{-1*}(z)[J - V^*(z)JV(z)]\mathfrak{A}_\kappa^{-1}(z) \geq 0.$$

The assertion of the Theorem now follows.

From this Theorem it follows that Weil's operators [4]

$$W_m(z) = \mathfrak{A}_m^{*-1}(z)J\mathfrak{A}_m^{-1}(z) = \begin{bmatrix} -r_m & s_m \\ s_m^* & -t_m \end{bmatrix} \tag{5.2}$$

are monotonically decreasing. For points z of the regions under consideration the strict inequality

$$r_m > 0 \tag{5.3}$$

is true. From (5.1) and (5.2) it follows that

$$w^*(z)r_m w(z) + is_m^* w(z) - iw^*(z)s_m \leq -t_m.$$

In view of (5.3), we write the last inequality in the form

$$[-iw^*(z)r_m^{\frac{1}{2}} + s_m^* r_m^{-\frac{1}{2}}][r_m^{-\frac{1}{2}}iw(z) + r_m^{-\frac{1}{2}}s_m] \leq s_m^* r_m^{-1} s_m - t_m. \tag{5.4}$$

It follows from (5.4) that

$$s_m^* r_m^{-1} s_m - t_m \geq 0.$$

Taking (5.4) into account we obtain

$$r_m^{\frac{1}{2}} iw(z) + r_m^{-\frac{1}{2}} s_m = V\rho_{\alpha m}^{\frac{1}{2}} \tag{5.5}$$

where

$$V^*V \leq E$$

and

$$\rho_{\alpha m} = s_m^* r_m^{-1} s_m - t_m. \tag{5.6}$$

We deduce from the formula (5.5) that

$$w(z) = c_m + \rho_{gm}^{\frac{1}{2}} \mathcal{U} \rho_{\alpha m}^{\frac{1}{2}}, \quad \mathcal{U}^* \mathcal{U} \leq E \tag{5.7}$$

where

$$c_m = ir_m^{-1} s_m, \quad \rho_{gm} = r_m^{-1}.$$

The set of operators admitting the representation (5.7) is called the Weyl operator disc with centre c_m, left radius ρ_{gm} and right radius ρ_{dm}.

It is clear from geometric considerations that the inequality (5.1) admits the following interpretation [17].

THEOREM 5.2. *For each fixed z the set of the operator functions $w(z)$ satisfying the inequality (5.1) determines an operator disc $\tilde{R}_m$ with centre $c_m = ir_m^{-1}s_m$, right radius $\rho_{\alpha m} = s_m^* r_m^{-1} s_m - t_m$ and left radius $\rho_{gm} = r_m^{-1}$.*

Let us investigate the behavior of Weyl discs as the parameter m increases. Suppose $w(z) \in \hat{R}_\kappa$ and $\kappa < m$. The operator functions $W_m(z)$ are monotonically decreasing with respect to m. Therefore the inequality

$$[-iw^*(z) \quad E] W_\kappa(z) \begin{bmatrix} iw(z) \\ E \end{bmatrix} \leq [-iw^*(z) \quad E] W_m(z) \begin{bmatrix} iw(z) \\ E \end{bmatrix}$$

holds. Hence $\tilde{R}_m \subset \tilde{R}_\kappa$, if $\kappa < m$. The following assertions are true:

1) The left and right radii of the Weyl discs decrease monotonically with respect to the parameter m and have the limit

$$\rho_d = \lim_{m \to \infty} \rho_{dm}, \quad \rho_g = \lim_{m \to \infty} \rho_{gm}$$

2) The family of centers c_m is bounded.

Using Orlov's Theorem [29] we obtain the assertions:

3) The ranks of the radii $\rho_d(z)$ and $\rho_g(z)$ of Weyl limiting discs do not depend on the choice of z.

If the parametrized family of centers c_m has a limit

$$c = \lim_{m \to \infty} c_m$$

then the set of operators

$$w(z) = c + \rho_g^{\frac{1}{2}} U \rho_d^{\frac{1}{2}}, \quad U^* U \leq E$$

is called Weyl's limit disc $\tilde{R}$. The limit disc $\tilde{R}$ belongs to every disc of $\tilde{R}_m$. If any of the limit radii ρ_d or ρ_g equal zero then Weyl's limit disc is a limit point. In the limit point case the interpolation problem has only one solution.

Let us say that the interpolation problem is determinate if it has only one solution, and that the problem is indeterminate if it has more than one solution.

6. DEGENERATE INTERPOLATION PROBLEMS AND THE METHOD OF REGULARIZATION

Rewrite identity (1.2) in the form

$$AS - SA^* = i(\Pi_1\Pi_2^* + \Pi_2\Pi_1^*) \tag{6.1}$$

The interpolation problem (1.1) is said to admit regularization if there is an operator $S_0 \in \{H; H\}$, which is bounded together with its inverse and an operator $\Pi_0 \in \{G_1; H\}$ such that

$$AS_0 - S_0A^* = i(\Pi_0\Pi_2^* + \Pi_2\Pi_0^*). \tag{6.2}$$

The method of regularization was worked out by Katsnel'son [6] for some classical interpolation problems.

THEOREM 6.1. *Suppose that the nonnegative operator S is bounded and satisfies the relation (6.2), the opertaor $A \in \mathcal{N}_0$ does not have zero as an eigenvalue, and $\Pi_2 g \neq 0$ if $g \neq 0$. If the corresponding interpolation problem admits regularization then there are an operator function $\sigma(t) \in \mathcal{E}$ and operators $\alpha = \alpha^*, \beta \geq 0$ such that the equalities (1.4), (1.6), (1.9) - (1.11) are true.*

PROOF: Consider the operators S_ϵ and $\Pi_{1,\epsilon}$ given by the formula

$$S_\epsilon = S + \epsilon S_0, \ \Pi_{1,\epsilon} = \Pi_1 + \epsilon\Pi_0, \ \epsilon > 0. \tag{6.3}$$

From (6.1) - (6.3), it follows that the operator identity

$$AS_\epsilon - S_\epsilon A^* = i(\Pi_{1,\epsilon}\Pi_2^* + \Pi_2\Pi_{1,\epsilon}^*) \tag{6.4}$$

holds. The regularized Fundamental Matrix Inequality

$$L_\epsilon(z) = \begin{bmatrix} S_\epsilon & B_\epsilon(z) \\ B_\epsilon^*(z) & C(z) \end{bmatrix} \geq 0, \ z \in \mathcal{D} \tag{6.5}$$

where

$$B_\epsilon(z) = (E - Az)^{-1}[\Pi_{1,\epsilon} - i\Pi_2 w(z)] \tag{6.6}$$

corresponds to the identity (6.4).

In view of (2.15), the following estimate

$$\|B_\epsilon(z)f\| = 0(1), \ \delta_1 \le |Arg\ z| \le \pi - \delta_1, \ |z| \ge \delta_2 \tag{6.7}$$

holds uniformly with respect to ϵ.

Suppose that $\{\epsilon_\kappa\}$ is the sequence of positive numbers such that $\epsilon_\kappa \to 0$ as $\kappa \to \infty$.

But according to Theorem 4.2, for each ϵ_κ the Regularized Fundamental Matrix Inequality (6.5) has the solution $w_\kappa(z)$. Taking into consideration (6.6) and (6.7), we obtain the estimate

$$\|\Pi_2 w_\kappa(z)\Pi_2^*\| = 0(1), \ \delta_1 \le |Arg\ z| \le \pi - \delta_1, \ |z| \ge \delta_2$$

which is uniform with respect to ϵ.

Since $\dim G_1 < \infty$ and $\Pi_2 g \ne 0$ if $g \ne 0$, we deduce from the last inequality that the estimate

$$\|w_\kappa(z)\| = 0(1), \ \delta_1 \le |arg\ z| \le \pi - \delta_1, \ |z| \ge \delta_2$$

holds uniformly with respect to κ. Therefore the family of operator functions $w_\kappa(z)$ is bounded on each compact set in $Im\ z \ne 0$. Choosing a converging subsequence from this family, we arrive at an operator function $w(z)$, analytic in $Im\ z \ne 0$ which becomes a solution of inequality (2.1). Then the operator function $w(z)$ has the representation (2.4) and according to Theorem 3.1 the operator function $\sigma(t)$ belongs to class $\mathcal{E}$. Taking limits, as $\epsilon_\kappa \to 0$, we obtain the equality (1.11). The theorem is proved.

We conclude that the interpolation problem has at least one solution if the conditions of Theorem 6.1 are fulfilled.

7. APPLICATIONS OF THE GENERAL THEORY

In this section we shall investigate some concrete interpolation problems by the operator method.

7.1 THE NEVANLINNA-PICK PROBLEM

We are given points

$$z_1, z_2, \cdots, z_n, \cdots \ (Im \ z_\kappa > 0, \ \kappa = 1, 2, \cdots)$$

and matrices of size $m \times m$:

$$w_1, w_2, \cdots, w_n, \cdots \ (\frac{w_\kappa - w_\kappa^*}{i} > 0, \ \kappa = 1, 2, \cdots)$$

Supposing that

$$dim \ G = 2m, \quad dim \ H = mn$$

we shall define the matrices

$$A = diag \ \{z_1 E_m, z_2 E_m, \cdots, z_n E_m\}; \tag{7.1}$$

$$\Pi = [\Pi_1 \quad \Pi_2], \ \Pi_1 = icol[w_1, w_2, \cdots, w_n], \ \Pi_2 = col[E_m, E_m, \cdots, E_m] \tag{7.2}$$

$$J = \begin{bmatrix} 0 & E_m \\ E_m & 0 \end{bmatrix}, \quad S = \Big[\frac{w_i - w_\kappa^*}{z_i - \bar{z}_\kappa} \Big]_{i,\kappa=1}^m . \tag{7.3}$$

The operator identity (1.2) immediately follows from the definition (7.1)-(7.3).

Now let us formulate the corresponding interpolation problem (7.1)-(7.3):

Describe the set of $(\sigma(t), \ \alpha, \ \beta)$, where $\sigma(t)$ is a monotonically increasing matrix function and α and β are matrices such that $\alpha = \alpha^*$, $\beta \geq 0$, the equality

$$w_\kappa = -\frac{\beta}{z_\kappa} - \alpha + \int_{-\infty}^{+\infty} \Big[z_\kappa(E - z_\kappa t)^{-1} + \frac{t}{1 + t^2} \Big] d\sigma(t), \ 1 \leq \kappa \leq n \tag{7.4}$$

holds and the integral

$$\tau = \int_{-\infty}^{+\infty} \frac{d\sigma(t)}{1 + t^2} < \infty \tag{7.5}$$

converges.

We remark that the formula (7.4) is equivalent to the representation $\Pi_1 = \tilde{\Pi}_1$. Since $A \in \mathcal{K}$ then according to Lemma 1.3 the representation

$$S = \left[\int_{-\infty}^{+\infty} \frac{d\sigma(t)}{(1 - z_\iota t)(1 - \bar{z}_\wedge t)} + \frac{\beta}{z_\iota \bar{z}_\wedge} \right]_{\iota,\wedge=1}^{n} \tag{7 6}$$

follows from (7.4).

Considering (2.4), (2.5) we give the following formulation of the associated interpolation problem:

Describe the set of matrix functions from the class $\mathcal{R}$ which satisfy the conditions

$$w\left(\frac{1}{z_\wedge}\right) = -w_\kappa.$$

This problem is the Nevanlinna-Pick problem for a finite number of nodes.

If $S > 0$ then from Theorem 4.2 and the formulas (4.16), (4.17) we have the general solution of the Nevanlinna-Pick problem.

Suppose $\Pi_1 = -\imath \Pi_2$ and consider the operator identity

$$AS_0 - S_0 A^* = 2\imath \Pi_2 \Pi_2^*. \tag{7.7}$$

From (7.7) it follows that

$$S_0 > 0.$$

Therefore the given problem admits regularization. It follows from Theorem 6.1 that for the solvability of Nevanlinna-Pick problem it is necessary and sufficient that

$$S \geq 0.$$

Plenty of literature is devoted to various methods of solution for Nevanlinna-Pick problems. Operator identities were used in the works [22]-[24], [3], [25].

7.2 THE HAMBURGER MOMENT PROBLEM

Let $S_0, S_1, \ldots, S_{2n-2}$ be given matrices of size $m \times m$. Supposing that dim $G = 2m$, dim $H = mn$, we introduce the matrices

$$
A = \begin{bmatrix} 0 & 0 & \cdots & 0 & 0 \\ E_m & 0 & \cdots & 0 & 0 \\ 0 & E_m & \cdots & 0 & 0 \\ \vdots & \vdots & \cdots & \vdots & \vdots \\ 0 & 0 & \cdots & E_m & 0 \end{bmatrix} ; \Pi_1 = -i \begin{bmatrix} 0 \\ S_0 \\ S_1 \\ \vdots \\ S_{n-2} \end{bmatrix} ; \Pi_2 = \begin{bmatrix} E_m \\ 0 \\ 0 \\ \vdots \\ 0 \end{bmatrix} ; J = \begin{bmatrix} 0 & E_m \\ E_m & 0 \end{bmatrix} \quad (7.8)
$$

$$
S = \begin{bmatrix} S_0 & S_1 & \cdots & S_{n-1} \\ S_1 & S_2 & \cdots & S_n \\ \vdots & \vdots & \cdots & \vdots \\ S_{n-1} & S_n & \cdots & S_{2n-2} \end{bmatrix} \quad (7.9)
$$

It is evident that the operator identity (1.2) and the identity

$$
(E - Az)^{-1}\Pi_2 = col[E_m, zE_m, \cdots, z^{n-1}E_m] \quad (7.10)
$$

are true. As the domains of values of A and Π_2 have only trivial intersection it follows from (2.12) that

$$
\beta = 0, \quad A(\Pi_1 - \tilde{\Pi}_1) = 0. \quad (7.11)
$$

The representation (1.11) and formulas (7.9), (7.10) lead to the following problem:

Describe the set of pairs $(\sigma(t), \alpha)$ where $\sigma(t)$ is a monotonically increasing matrix function and $\alpha = \alpha^*$ is a Hermitian matrix such that

$$
\int_{-\infty}^{+\infty} (1 + t^{2n-2})d\sigma(t) < \infty \quad (7.12)
$$

and

$$
S_\kappa = \int_{-\infty}^{+\infty} t^\kappa d\sigma(t), \ 0 \le \kappa \le 2n - 2; \quad \alpha = \int_{-\infty}^{+\infty} \frac{t}{1 + t^2} d\sigma(t). \quad (7.13)
$$

This amounts to a formulation of the Hamburger moment problem for a finite number of interpolation values. It follows from Theorem 3.4 and (7.11) that

$$\alpha = \int_{-\infty}^{+\infty} \frac{t}{1+t^2} d\sigma(t); \quad S_\kappa = \int_{-\infty}^{+\infty} t^\kappa d\sigma(t), \quad 0 \le \kappa \le 2n-3. \tag{7.14}$$

In view of (7.11) and (7.14), we see that the formula (2.4) takes the form

$$w(z) = \int_{-\infty}^{+\infty} \frac{d\sigma(t)}{t-z}.$$

From the transformed inequality (3.6) and equations (3.5), (3.10), (3.11) we deduce the inequality

$$S_{2n-2} \ge \int_{-\infty}^{+\infty} t^{2n-2} d\sigma(t). \tag{7.15}$$

The solution of the Hamburger moment problem demands a modification of our previous approach. The adjustment is that instead of $S = \tilde{S}$ one has only the weaker equality

$$A(S - \tilde{S}) = 0 \tag{7.16}$$

The inequality (2.1) is equivalent to the following modified Hamburger moment problem.

Describe the set of all pairs $(\sigma(t), \alpha)$, where $\sigma(t)$ is a monotonically increasing matrix function and $\alpha = \alpha^*$ is a Hermitian matrix which satisfy (7.12), (7.13) and (7.16), and where $S_0, S_1, \cdots, S_{2n-2}$ are given matrices of size $m \times m$.

If $S > 0$ then Theorem 4.2 can be applied and the formulas (4.16), (4.17) give the general solution of modified Hamburger moment problem for a finite number of interpolation values. It follows from Theorem 6.1 that the nonnegativity of the information block $S \ge 0$ is a necessary and sufficient condition for solvability of the Hamburger moment problem.

Remark. For the Hamburger problem only the modification of the original problem is equivalent to the inequality (2.1). In this case the inequality (2.1) is both the basis for the method of solution and an aid to the formulation of the problem itself. We assume that such an approach (the modification of original problem dictated by the inequality (2.1)) is useful on other occasions when there is no equivalence of the interpolation problem and inequality (2.1).

7.3 OPERATORS WITH W-DIFFERENCE KERNELS

Designate the space of vector functions

$$\vec{f}(t) = col[f_1(t), f_2(t), \cdots, f_n(t)]$$

with the norm

$$\|\vec{f}\|^2 = \int_0^\omega (\sum_{\kappa=1}^n |f_\kappa(t)|^2) dt.$$

by $L_n^2(0, \omega)$. Here we shall consider the case when $H = L_n^2(0, \omega)$ and the space G_1 is given by a constant vector

$$\vec{g} = col[g_1, g_2, \cdots, g_n].$$

We suppose that

$$W = diag \{w_1, w_2, \cdots, w_n\}.$$

Let us say that the operator

$$S\vec{f} = \frac{d}{dx} \int_0^\omega S(x, t)\vec{f}(t) dt \tag{7.17}$$

has W-difference kernel if the kernel $S(x, t)$ has the form

$$S(x, t) = \{S_{\ell, m}(\omega_m x - \omega_\ell t)\}_{\ell, m=1}^n; \ S_{\ell, m}(u) \in L^2(-\omega_\ell \omega; \omega_m \omega). \tag{7.18}$$

If $\omega_1 = \omega_2 = \cdots = \omega_n$ then S is the operator with difference kernel, i.e. $S(x, t) = S(x - t)$. If $\omega_1 = \omega_2 = \cdots = \omega_p = a$, $\omega_{p+1} = \omega_{p+2} = \cdots = \omega_n = -a$ then the kernel $S(x, t)$ has the form

$$S(x, t) = \begin{bmatrix} S_{11}(x - t) & S_{12}(x + t) \\ S_{21}(x + t) & S_{22}(x - t) \end{bmatrix}.$$

In addition, we define

$$A\vec{f} = iW \int_0^x \vec{f}(t) dt, \ M(x) = WS(x, 0), \ N(x) = -WS(0, x).$$

By direct calculation we get the operator identity

$$(AS - SA^*)\vec{f} = i \int_0^\omega [M(x) + N(t)]\vec{f}(t)dt. \qquad (7.19)$$

We suppose that the operator S is selfadjoint, i.e.

$$N(x) = M^*(x).$$

Let us introduce the operators acting from G_1 into $L_n^2(0, \omega)$ given by

$$\Pi_1\vec{g} = M(x)\vec{g}, \quad \Pi_2\vec{g} = \vec{g}.$$

Then the identity (7.19) can be rewritten in the form

$$AS - SA^* = i\Pi J\Pi^*. \qquad (7.20)$$

Taking into consideration the equality

$$(E - At)^{-1}\Pi_2 = exp\,(iWxt)\Pi_2$$

we obtain that the integral representations $S = \tilde{S}$ and $\Pi = \tilde{\Pi}$ are equivalent to the representation

$$S(x,t) = S(0,t) + \frac{\partial}{\partial t} \int_{-\infty}^{+\infty} (Wu)^{-1}(e^{ixWu} - E)[d\sigma(u)](e^{-iWtu} - E)(Wu)^{-1} \qquad (7.21)$$

for $0 \le x, t \le w$.

It is evident that $A \in \mathcal{N}_0$, $\Pi_2 g \ne 0$ if $g \ne 0$ and the domains of values of A and Π_2 have only trivial intersection.

THEOREM 7.1. *Suppose that the operator S in (7.17), (7.18) is bounded. Then the set of $\sigma(t) \in \mathcal{E}_t$ which give the representation (7.21) coincides with the set of $\sigma(t)$ generated by the class of $w(z)$ satisfying the FMI of the problem under consideration via the formula*

$$w(z) = \beta z + \alpha + \int_{-\infty}^{+\infty} \left(\frac{1}{t-z} - \frac{t}{1+t^2}\right) d\sigma(t).$$

If the operator S is invertible then the solutions of inequality (2.1) are given by Formula 4.9.

THEOREM 7.2. *If the operator S in (7.17), (7.18) is bounded then the operator function $\sigma(t) \in \mathcal{E}_t$ satisfying the equality (7.21) exists if and only if*

$$S \geq 0.$$

There are classical works by Krein [28] devoted to the problem of representation of operator kernels (7.17), (7.18) if $n = 1$ and the function S has the form $S(x,t) = S(x-t)$ with S differentiable. Under the same conditions the operator (7.17), (7.18) was investigated in Potapov's and Kovalishina's works [18] and Katsnel'son's works [6]. For the case of invertible operators of the type (7.17), (7.18) with $n = 1$ the formula (7.21) was deduced by A. Sakhnovich [30].

If S is the identity operator E, then (7.21) and formula (1.7) can be rewritten in the form

$$A - A^* = i\Pi j \Pi^*,$$

$$E = \int_{-\infty}^{+\infty} (E - At)^{-1}\Pi_2[d\sigma(t)]\Pi_2^*(E - A^*t)^{-1} \quad (F = 0).$$

We note that the interpolation problem studied by Louis de Branges [9] is connected with the case $S = E$.

REFERENCES

1. V.M. Adamjan, D.Z. Arov, M.G. Krein. Analytic properties of Schmidt pairs for a Hankel operator and the generalized Schur-Takagi problem, *Math. USSR - Sb.* 15, 31-73, 1971.

2. M.G. Krein, H. Langer. Über einige Fortsetzung probleme. *Math. Nachr.* 77, 187-236, 1977.

3. A.A. Nudel'man, On a new problem of moment type. *Dokl. Akad. Nauk SSSR* 233 (1977), 792-795; *Soviet Math. Dokl.* 18 (1977), 507-510.

4. I.V. Kovalishina, V.P. Potapov, Indefinite metric in the Nevanlinna-Pick problem. *Dokl. Akad. Nauk Armyan. SSR, Ser. Mat.*, 59 (1974), 17-22; (Translation) *Collected Papers of V.P. Potapov*, p.p. 33-40, Hokkaido Univ., 1982, Sapporo.

5. *J*-expansive matrix-functions in the Caratheodory problem, *Dokl. Akad. Nauk Armyan. SSR, Ser. Mat.*, 59 (1974), 129-135.

6. V.E. Katsnel'son, Methods of J-theory in continuous interpolation problems of analysis. Part 1. Hokkaido Univ. 1985, Sapporo.

7. I.S. Iohvidov, M.G. Krein, H. Langer, *Introduction to the Spectral Theory of Operators in Spaces with an Indefinite Metric.* Akademic-Verlag Berlin, Band 9, 1982.

8. C. Foias, A. Frazho, *The Commutant Lifting Approach to Interpolation Problems*, Birkhäuser-Verlag, 1990.

9. L. de Branges, *Hilbert Spaces of Entire Functions*, 1968, Prentice-Hall. - 326 p.

10. J. Ball, I. Gohberg, L. Rodman, *Interpolation of Rational Matrix Functions.* Birkhäuser-Verlag, 1990.

11. P. Dewilde, A Course on the Algebraic Schur and Nevanlinna-Pick Interpolation Problems. *Lectures and Tutorials Presented at the International Workshop on Algorithums and Parallel VLSI Architectures*, Abbaye des Premontres, Pont-a-Mous-son, France, June 10-16, 1990.

12. H. Nijmeijer, J.M. Schumacher (ed.), *Three Decades of Mathematical System Theory*, Springer-Verlag, 1989.-562 p.

13. V.P. Potapov, The multiplicative structure of J-contractive matrix functions. *Trudy Moskov. Mat. Obshch.*, 4 (1955), 125-236; (Translation) *Amer. Math. Soc. Transl. (2)* 15 (1960), 131-243.

14. A.V. Efimov, V.P. Potapov. J-expanding matrix functions and their role in the analytical theory of electrical circuits. *Uspehi Mat. Nauk*, 28 (1973), no 1 (169) 65-130; (Translation) *Russian Math. Survey*, 28:1 (1973), 69-140.

15. V.P. Potapov, General theorems on the structure and splitting-off of elementary factors of analytic matrix-functions. *Dokl. Akad. Nauk. Armyan. SSR, Ser. Mat.*, 48 (1969), 257-262; (Translation) *Collected papers of V.P. Potapov*, pp. 23-32, Hokkaido Univ., 1982, Sapporo.

16. V.P. Potapov, Fundamental facts of the theory of J-contractive matrix-functions. *Proc. All Union Conf. on Theory of Functions*, pp. 1979-181, 1971, Khar'kov.

17. I.V. Kovalishina, V.P. Potapov. The radii of Weyl disc in the tangential Nevanlinna-Pick problem, in *Theory of Operators in Function Spaces and its Applications*, pp. 25-49, Naukova Dumka, 1981, Kiev; (Translation) *Collected Papers of V.P. Potapov*, pp. 67-99, Hokkaido Univ., 1982, Sapporo.

18. I.V. Kovalishina, V.P. Potapov. Integral representation of Hermitian positive functions. Deposited to VINITI, no. 2984-81, 1981; (Translation) Hokkaido Univ., 1982, Sapporo.

19. V.K. Dubovoi, Indefinite metric in the interpolation Schur problem for analitic functions. *Teor Funktsii Funktsional Anal. i Prilozhen.*, I - 37 (1982), 14-26; II - 38 (1982), 32-40; III - 41 (1984), 55-64; IV - 42 (1984), 46-57.

20. L.A. Sakhnovich, On similarity of linear operators. *Siberian Math. J.* 13 (1972), 604-515.

21. T.S. Ivanchenko, L.A. Sakhnovich, Operator approach to the investgation of interpolation problems. Deposited to Ukr. NIINTI, no. 701, Uk-85, 1985.

22. T.S. Ivanchenko, L.A. Sakhnovich, Operator identities in the theory of interpolation problems. *Izv. Akad. Nauk Armyan. SSR. Ser. Mat.*, XXII, 3 (1987), 298-308.

23. T.S. Ivanchenko, L.A. Sakhnovich, Operator approach to Potapov scheme. *Ukrain Math. J.* 39 no. 5 (1987), 573-578.

24. A.A. Nudel'man, P.A. Shvartsman, On the existence of solutions of certain operator inequalities, *Siberian Math J.* 16 (1975) 431-439. Plenum Publishing Co., New York.

25. L.A. Sakhnovich, Equations with a difference kernel on a finite interval. *Russian Math. Surv.* 35:4 (1980) 81-152.

26. L.A. Sakhnovich, Factorization problems and operator identities. *Russian Math. Surv.* 41:1 (1986) 1-64.

27. M.G. Krein, Sur le probleme du prolongement des fonctions hermitiennes positivees et continues. *Dokl. Akad. Nauk SSSR*, 26 (1940), 17-22.

28. S.A. Orlov, Nested matrix disks, analytically depending on a parameter, and theorems on invariance of ranks of radii of limiting disks. *Izv. Akad. Nauk SSSR, Ser. Mat.*, 40 (1976), 593-644; (Translation) *Math USSR - Izv.*, 10 (1976), 565-613.

29. A.L. Sakhnovich, On the extension of Toeplitz matrices and their continuous analogues. Dissertation, Khar'kov (1982).

L.A. Sakhnovich
Pr. Dobrovolskogo 154 ap. 199
Odessa 270111
Ukraine

Operator Theory:
Advances and Applications, Vol. 72
© 1994 Birkhäuser Verlag Basel

DESCRIPTION OF A CLASS OF FUNCTIONS WHICH ADMIT AN APPROXIMATION BY RATIONAL FUNCTIONS WITH PREASSIGNED POLES I.

Victor E. Katsnelson

In classical complex function theory there are problems for scalar functions which can be better understood and studied in the framework of J-contractive analytic matrix functions. This point of view was initiated by V.P.Potapov who created the theory of J-contractive analytic matrix functions (originally with other motivations). Propagating this viewpoint he stressed the fact of the same nature that some problems of real analysis can be better understood by using methods of complex function theory. The validity of this point of view was confirmed by V.P.Potapov's own work, and investigations of his followers, on extrapolation and interpolation problems for functions belonging to special classes (of I.Schur, C.Caratheodory, M.G.Krein, etc.) and on approximation problems for scalar functions. The present article, which is devoted to the solution of a problem for scalar analytic functions which was posed 25 years ago by G.Ts.Tumarkin, can be regarded as further evidence of the fruitfulness of the ideology of V.P.Potapov.

Notation: $\mathbb{N}$ is the set of natural numbers, $\mathbb{R}$ is the real axis, $\mathbb{C}$ is the complex plane, $\overline{\mathbb{C}} := \mathbb{C} \cup \{\infty\}$, $\mathbb{D}_+ := \{z \in \mathbb{C} : |z| < 1\}$, $\mathbb{D}_- := \{z \in \overline{\mathbb{C}} : 1 < |z| \le \infty\}$, $\mathbb{T} := \{t \in \mathbb{C} : |t| = 1\}$, $m(dt)$ is the normalized Lebesgue measure on $\mathbb{T}$. We use the notation "a.e." as an abbreviation for "almost everywhere" with respect to the Lebesgue measure $m(dt)$. If f is a function defined on a set K and L is a subset of K, then $Rstr._L f$ stands for the restriction of f to L.

The work was supported by the Minerva Foundation

1. Introduction.

Let $\mathcal{T} = [a_{kj}]$ be a given table of points from $\overline{\mathbb{C}}$:

$$
\mathcal{T} = \begin{bmatrix}
a_{11}, & a_{12}, & \ldots, & a_{1n_1} \\
a_{21}, & a_{22}, & \ldots, & a_{2n_2} \\
\cdot & \cdot & \cdot & \cdot \\
a_{k1} & a_{k2} & \ldots & a_{kn_k} \\
\cdot & \cdot & \cdot & \cdot
\end{bmatrix}
\tag{1.1}
$$

$a_{kj} \in \overline{\mathbb{C}}$, $1 \leq j \leq n_k$, $n_k < \infty$, $k = 1, 2, 3, \ldots$ which are subject to only one restriction:

$$
|a_{kj}| \neq 1, \qquad 1 \leq j \leq n_k, \ 1 \leq k < \infty
\tag{1.2}
$$

Definition 1.1. Let $\mathcal{T}$ be a table (1.1) of points from $\overline{\mathbb{C}}$. A sequence $\{R_k\}_{k \in \mathbb{N}}$ of rational functions is *subordinated to the given table* $\mathcal{T}$ if for each k the poles of the function R_k are contained (counting multiplicities) in the set $\{a_{k1}, a_{k2}, \ldots, a_{kn_k}\}$ of the points of the k-th row of the table $\mathcal{T}$. The set of all sequences $\{R_k\}_{k \in \mathbb{N}}$ of rational functions which are subordinated to the given table $\mathcal{T} = [a_{kj}]$ is denoted by $\mathcal{R}(\mathcal{T})$ or by $\mathcal{R}([a_{kj}])$.

If all points a_{kj} in the k-th row are distinct and $a_{kj} \neq \infty$ ($1 \leq j \leq n_k$), then the function R_k is representable in the form

$$
R_k(t) = c_{k0} + \sum_{1 \leq j \leq n_k} \frac{c_{kj}}{t - a_{kj}},
\tag{1.3}
$$

where c_{kj}, $0 \leq j \leq n_k$, are complex numbers. Thus the points a_{kj} are prescribed, and the coefficients c_{kj} remain to be chosen. If some numbers a of the k-th row occur more than once, this means, that the function R_k can have a pole at a with multiplicity not exceeding the multiplicity of the point a. We allow also $a_{kj} = \infty$ for some j.

Let $\mathcal{M}$ be a metric space whose elements are functions defined on $\mathbb{T}$. J.L. Walsh [W] posed the following approximation problem: Let $\mathcal{T} = [a_{kj}]$ be a table (1.1). Give necessary and sufficient conditions for $f \in \mathcal{M}$ to be approximated (with respect to the topology of the space $\mathcal{M}$) by a sequence $\{R_k\}_{k \in \mathbb{N}}$ from $\mathcal{R}(\mathcal{T})$. In the papers [Tum5], [Tum6], [Tum7] of G.Ts. Tumarkin this approximation problem was considered in the case where the metric space $\mathcal{M}$ of function on $\mathbb{T}$ is a weighted space L_w^p. Namely, let w be a Borel function defined on $\mathbb{T}$, $w : \mathbb{T} \to [0, +\infty]$. This function w is said to be *a weight function.* Let p be a given positive number, $0 < p < \infty$. Let us define the metric space L_w^p. Its elements are Borel functions x on $\mathbb{T}$ which satisfy the condition

$\int_{\mathbb{T}} |x(t)|^p w(t) m(dt) < \infty$. If $p \in [1, \infty)$, the distance $\rho_{L_w^p}(x, y)$ between two elements x and y is defined as

$$\rho_{L_w^p}(x, y) := \{ \int_{\mathbb{T}} |x(t) - y(t)|^p w(t) m(dt) \}^{\frac{1}{p}} .$$

If $p \in (0, 1)$ the distance $\rho_{L_w^p}(x, y)$ is defined as

$$\rho_{L_w^p}(x, y) := \int_{\mathbb{T}} |\dot{x}(t) - y(t)|^p w(t) m(dt).$$

We suppose that the weight function $w(t)$ satisfies the following condition:

i). *The function $w(t)$ is $m(dt)$-summable, i.e.,*

$$\int_{\mathbb{T}} w(t) m(dt) < \infty. \tag{1.4}$$

ii). *The logarithmic integral converges, i.e.,*

$$\int_{\mathbb{T}} [\ell n \ w(t)] m(dt) > -\infty. \tag{1.5}$$

(Condition (1.5) is also called *the Szegö condition*).

Definition 1.2.   By $\mathcal{W}$ we denote the set of all weight function w which satisfy the conditions (1.4) and (1.5).

Let us associate two numbers Cap_k^+ and Cap_k^- with the k-th row of the table $\mathcal{T} = [a_{kj}]$:

$$Cap_k^+ := \sum_{j : a_{kj} \in \mathbb{D}_+} (1 - |a_{kj}|), \qquad Cap_k^- := \sum_{j : a_{kj} \in \mathbb{D}_-} (1 - \frac{1}{|a_{kj}|}). \tag{1.6}$$

(The notation $Cap_k^{\pm}$ is adopted from [Nik 1; Lecture II] and [Nik-Gr]. See also the book [Nik 2; Lecture 2, item 3], where the notion of *the capacity of rational function* was introduced).

The properties of a function f which is the limit (with respect to the L_w^p topology) of a sequence $\{R_k\} \in \mathcal{R}([a_{kj}])$ depend essentially on the behavior of the sequences Cap_k^+ and Cap_k^-. There are three essentially different possibilities:

α).
$$\underline{\lim}_{k\to\infty} Cap_k^+ < \infty, \qquad \underline{\lim}_{k\to\infty} Cap_k^- < \infty; \tag{1.7}$$

β).
$$\begin{cases} \beta' : \underline{\lim}_{k\to\infty} Cap_k^+ < \infty, \qquad \lim_{k\to\infty} Cap_k^- = \infty, \\[1mm] \qquad \text{or (what is in fact the same)} \\[1mm] \beta'' : \lim_{k\to\infty} Cap_k^+ = \infty, \qquad \overline{\lim}_{k\to\infty} Cap_k^- < \infty; \end{cases} \tag{1.8}$$

γ).
$$\lim_{k\to\infty} Cap_k^+ = \infty, \qquad \lim_{k\to\infty} Cap_k^- = \infty; \tag{1.9}$$

The case γ) is the simplest one to investigate. If the table $\mathcal{T} = [a_{kj}]$ satisfies condition (1.9), then each function f from L_w^p is the limit of a sequence $\{R_k\}$ from $\mathcal{R}([a_{kj}])$. The case β) is more difficult. It was completely investigated in the papers [Tum5] and [Tum6] of G. Ts. Tumarkin. There he obtained necessary and sufficient conditions for a given function $f \in L_w^p$ to admit an approximation by a sequence $\{R_k\} \in \mathcal{R}([a_{kj}])$ for a given table $\mathcal{T} = [a_{kj}]$ under the constraint (1.8).

The case α) is the most difficult one. It has not been investigated completely up to now. In [Tum7], G. Ts. Tumarkin obtained necessary conditions for a given function $f \in L_w^p$ to be the limit (with respect to the L_w^p-topology) of a sequence $\{R_k\} \in \mathcal{R}([a_{kj}])$ for a given table $\mathcal{T} = [a_{kj}]$ under the condition (1.7). In the same paper he also proved that these necessary conditions are sufficient in the special case $p \in [1, \infty)$, $w(t) \equiv 1$. The problem of describing the class of functions $f \in L_w^p$ which admit an approximation by a sequence $\{R_k\} \in \mathcal{R}([a_{kj}])$, under constraint (1.7), was posed as an open problem in [Tum7].

The investigations of G. Ts. Tumarkin are summarized in [Tum8].

Given a table $\mathcal{T} = [a_{kj}]$, let us associate with it two sequences of Blaschke products $\{B_k^+\}$ and $\{B_k^-\}$:

$$B_k^\pm(z) := \prod_{j:a_{kj}\in\mathbb{D}_\pm} \frac{a_{kj} - z}{1 - \bar{a}_{kj}z} \cdot \frac{|a_{kj}|}{a_{kj}}, \qquad (k \in \mathbb{N}). \tag{1.10}$$

(If, for example, the set $\{j : a_{kj} \in \mathbb{D}_-\}$ is empty, then, by definition, $B_k^-(z) \equiv 1$.) Let

$$V_{\mathcal{T}}^\pm(z) := \overline{\lim}_{k\to\infty} \ell n |B_k^\pm(z)|, \qquad (z \in \mathbb{D}_\pm). \tag{1.11}$$

It turns out that the functions $V_{\mathcal{T}}^\pm$ are the logarithmes of the moduli of contractive holomorphic functions: there exist a function $C_{\mathcal{T}}^+$ holomorphic in $\mathbb{D}_+$ and a function $C_{\mathcal{T}}^-$ holomorphic in $\mathbb{D}_-$ with

$$|C_{\mathcal{T}}^+(z)| \leq 1 \quad (z \in \mathbb{D}_+) \qquad \text{and} \qquad |C_{\mathcal{T}}^-(z)| \leq 1 \quad (z \in \mathbb{D}_-)$$

such that

$$V_{\mathcal{J}}^{\pm}(z) = \ell n|C_{\mathcal{J}}^{\pm}(z)| \qquad (z \in \mathbf{D}_{\pm}).$$

The functions $C_{\mathcal{J}}^{\pm}$ were defined in [Tum7] as the best analytic majorants of the families $\{\mathcal{B}^{\pm}\}$ of all possible limit functions of all subsequences $\{B_{k_l}^{\pm}(z)\}$ which converge locally uniformly in $\mathbf{D}_+$ and $\mathbf{D}_-$, respectively. (For more information on the existence and properties of the analytic majorant of a family of Blaschke products, see [Tum1], [Tum3], [Tum4]). Condition (1.7) is equivalent to the condition

$$C_{\mathcal{J}}^{+} \not\equiv 0 \qquad \text{and} \qquad C_{\mathcal{J}}^{-} \not\equiv 0 \tag{1.12}$$

We assume in the sequel that condition (1.7) (or, what is the same, condition (1.12)) is satisfied.

Let

$$C_{\mathcal{J}}^{+}(z) = I_{\mathcal{J}}^{+}(z)E_{\mathcal{J}}^{+}(z) \quad (z \in \mathbf{D}_+), \qquad \text{and} \qquad C_{\mathcal{J}}^{-}(z) = I_{\mathcal{J}}^{-}(z)E_{\mathcal{J}}^{-}(z) \quad (z \in \mathbf{D}_-)$$

be the inner-outer factorizations of the functions $C_{\mathcal{J}}^{+}$ and $C_{\mathcal{J}}^{-}$, respectively. For us only the inner functions $I_{\mathcal{J}}^{+}$ and $I_{\mathcal{J}}^{-}$ are interesting. They are determined by the given table $\mathcal{J} = [a_{kj}]$ only and completely characterize this table.

Definition 1.3. The inner functions $I_{\mathcal{J}}^{+}$ and $I_{\mathcal{J}}^{-}$ are said to be *the plus-denominator and the minus-denominator corresponding to the table* $\mathcal{J} = [a_{kj}]$, respectively.

Two *different* tables which have the same plus- and minus-denominators are considered as *equivalent* tables.

Remark 1.1. The usage of the terminology "plus-*denominator* and minus-*denominator* corresponding to the table $[a_{kj}]$" may seem to be rather strange. We use this terminology to emphasize the close relationship between the functions $I_{\mathcal{J}}^{+}$, $I_{\mathcal{J}}^{-}$ and the *denominators* which arise in Arov's paper [Ar2] on Darlington synthesis. D.Z. Arov designated his "*denominators*" by b_1, b_2. (The formulas (4.14) in [Ar2].) See also formulas (1) in [Ar4], (1.1) and (1.3) in [Ar5] and [4] in [Ar6], where the "*denominators*" b_1 and b_2 appear in connection with the so-called generalized Schur-Nevanlinna-Pick problem.

Remark 1.2. The above construction of the inner functions $I_{\mathcal{J}}^{\pm}$ (the denominators corresponding to the table $\mathcal{J}$) is based on results of G.Ts. Tumarkin concerning the existence of a least analytic majorant for a given sequence of Blaschke products. G.Ts. Tumarkin obtained these results by methods of classical complex function theory. Tumarkin's construction can also be interpreted in the language of spectral function theory. (For an exposition of spectral function theory, see the book of N.K. Nikol'skiĭ [Nik 2].) Let us consider the Hilbert space $H^2(\mathbb{D}_+)$ (the so-called Hardy space) and the unilateral shift operator $\mathbf{S}$ acting on $H^2(\mathbb{D}_+)$: $(\mathbf{S}h)(z) := zh(z)$ $(h \in H^2(\mathbb{D}_+))$. By Beurling's theorem, there exists one-to-one correspondence between inner functions and (nonzero) $\mathbf{S}$-invariant closed subspaces of the space $H^2(\mathbb{D}_+)$. The multiplication operator by an inner function is an isometric operator which acts on the space $H^2(\mathbb{D}_+)$ and commutes with the shift operator $\mathbf{S}$. Let $\{I_k^+\}$ be a given sequence of inner functions in $\mathbb{D}_+$ which does not converge to zero. This sequence has at least one convergent subsequence with a nonzero limit function. (In general, this limit function is $\underline{\text{not}}$ inner; it is, however, a contractive function.) The operator of multiplication by any limit function commutes with the shift $\mathbf{S}$. Thus, its image is an $\mathbf{S}$-invariant subspace. Let $\{\mathcal{L}^+\}$ be a set of $\mathbf{S}$-invariant subspaces which correspond to the set $\{\mathfrak{I}^+\}$ of all limit functions of the sequence $\{I_k^+\}$. The closed linear span of the set $\{\mathcal{L}^+\}$ is a closed $\mathbf{S}$-invariant subspace of the space $H^2(\mathbb{D}_+)$. (As the sequence $\{I_k^+\}$ does not converge to zero, this is a nonzero subspace.) By Beurling's theorem, this subspace has the form $I^+ H^2(\mathbb{D}_+)$ where I^+ is an inner function in $\mathbb{D}_+$. (It is possible that $I^+ \equiv 1$.) This function I^+ is the least analytic majorant of the sequence $\{I_k^+\}$. For a sequence $\{I_k^-\}$ of inner function in $\mathbb{D}_-$ the situation is similar. The spectral interpretation of Tumarkin's results, and further developments have been carried out by M.B. Gribov and N.K. Nikol'skiĭ [Gr-Nik]. In general, the multiplicative decomposition of the denominators $I_{\mathfrak{J}}^{\pm}$ contains both Blaschke components $B_{\mathfrak{J}}^{\pm}$ and singular components $S_{\mathfrak{J}}^{\pm}$:

$$I_{\mathfrak{J}}^+ = B_{\mathfrak{J}}^+ S_{\mathfrak{J}}^+, \qquad I_{\mathfrak{J}}^- = B_{\mathfrak{J}}^- S_{\mathfrak{J}}^- \tag{1.13}$$

However, it can happen that $I_{\mathfrak{J}}^+ = B_{\mathfrak{J}}^+$ (i.e. $S_{\mathfrak{J}}^+ \equiv 1$), or that $I_{\mathfrak{J}}^+ = S_{\mathfrak{J}}^+$ (i.e. $B_{\mathfrak{J}}^+ \equiv 1$).

Example 1. Let $\{a_k\}_{k \in \mathbb{N}}$ be a given sequence of comlex numbers subject to the restriction $|a_k| \neq 1$ $(k \in \mathbb{N})$, and let

$$\sum_{k:\ a_k \in \mathbb{D}_+} (1 - |a_k|) + \sum_{k:\ a_k \in \mathbb{D}_-} (1 - |a_k|^{-1}) < \infty$$

Let the table $\mathfrak{J}$ have the form

$$\begin{bmatrix} a_1 & & & \\ a_1, & a_2 & & \\ \cdot & \cdot & \cdot & \\ a_1, & a_2, & \ldots, & a_k \\ \cdot & \cdot & & \cdot \end{bmatrix}$$

Then the denominators $I_{\mathcal{J}}^{+}$ and $I_{\mathcal{J}}^{-}$ are both Blaschke products:

$$I_{\mathcal{J}}^{\pm}(z) = \prod_{k:\ a_k \in \mathbb{D}_{\pm}} \frac{a_k - z}{1 - \bar{a}_k z} \cdot \frac{|a_k|}{a_k}.$$

Example 2. Let α be a given positive number < 1. Let the table $\mathcal{J}$ have the form

$$\begin{bmatrix}
1 - \alpha \\[2mm]
1 - \frac{\alpha}{2} & 1 - \frac{\alpha}{2} \\[2mm]
1 - \frac{\alpha}{3} & 1 - \frac{\alpha}{3} & 1 - \frac{\alpha}{3} \\[2mm]
\cdot & \cdot & \cdot \\[2mm]
1 - \frac{\alpha}{k} & 1 - \frac{\alpha}{k} & \cdots & 1 - \frac{\alpha}{k} \\[2mm]
\cdot & \cdot & \cdot & \cdot
\end{bmatrix} \tag{1.14}$$

Then condition (1.7) is satisfied, since for all k the numbers $Cap_k^{\pm}$ defined in (1.6) are equal to $Cap_k^{+} \equiv \alpha$, $Cap_k^{-} \equiv 0$ respectively. It is not difficult to calculate the denominators for the table $\mathcal{J}$, (1.14):

$$I_{\mathcal{J}}^{+}(z) = exp\left\{-\alpha\frac{1+z}{1-z}\right\}, \qquad I_{\mathcal{J}}^{-}(z) = 1.$$

From results of the papers [Tum1], [Tum4] of G.Ts. Tumarkin we obtain

Criterion for the table to generate pure Blaschke denominators. *The following conditions are necessary and sufficient for the given table $\mathcal{J} = [a_{kj}]$ to generate pure Blaschke denominators:*

$$\lim_{r\uparrow 1}\left(\sup_{k} \sum_{j:\ r<|a_{kj}|<1} (1 - |a_{kj}|)\right) = 0,$$

and

$$\lim_{r\downarrow 1}\left(\sup_{k} \sum_{j:\ 1<|a_{kj}|<r} (1 - |a_{kj}|^{-1})\right) = 0.$$

In this paper we give the solution of the following problem: *Let I^{+} and I^{-} be given inner functions in $\mathbb{D}_{+}$ and $\mathbb{D}_{-}$, respectively. What are the necessary and sufficient conditions for the function $f \in L_w^p$ to be the limit of the sequence $\{R_k\} \in \mathcal{R}(\mathcal{J})$, where $\mathcal{J} = [a_{kj}]$ is a table with I^{+} and I^{-} as its plus- and minus denominator?*

We consider the *general case* here: $p \in (0, \infty)$, $w \in \mathcal{W}$. In the first part of this paper we restrict ourselves to the particular case where the given inner functions I^+, I^- are pure Blaschke products. This means that the singular factors $S_{\mathcal{J}}^+$ and $S_{\mathcal{J}}^-$ of the multiplicative decompositions (1.13) are trivial: $S_{\mathcal{J}}^+ \equiv 1$, $S_{\mathcal{J}}^- \equiv 1$.

Our proofs are not very long but require a number of deep facts, namely Frostman's theorem from the theory of value distribution of holomorphic functions (more precisely, its generalization due to W. Rudin, (see [Rud1]) and (independently) S.A. Vinogradov [Vin]), and a result of D.Z. Arov [Ar3] on the approximation of a pseudocontinuable function which is bounded on $\mathbb{T}$. This theorem of D.Z. Arov itself relies on two deep facts, namely the possibility of constructing a Darlington realization for each holomorphic matrix-valued function in $\mathbb{D}_+$ which is contractive in $\mathbb{D}_+$ and pseudocontinuable (this was proved by D.Z. Arov, [Ar1], [Ar2] and by R.G. Douglas and J.W. Helton [DH] independently), and a theorem due to V.P. Potapov (see [Pot1], [Pot2], [Gin]) on the multiplicative decomposition of a matrix-valued function which is analytic and contractive in $\mathbb{D}_+$.

It is surprising that the proof of the theorem which is formulated in a "purely scalar setting" (at least it is not obviously related to matrix- valued functions) uses deep facts from the theory of analytic matrix-valued functions.

2. The Class $PCNM$ of Pseudocontinuable Functions.

If G is some subset of $\overline{\mathbb{C}}$, then $G^{\#}$ denotes that subset of $\overline{\mathbb{C}}$ symmetric to G with respect to the unit circle, i.e.

$$G^{\#} := \{z \in \overline{\mathbb{C}} : \frac{1}{\overline{z}} \in G\}. \tag{2.1}$$

If $f : G \to \mathbb{C}$, then $f^{\#}$ is that complex-valued function which is defined on $G^{\#}$ by the rule

$$f^{\#}(z) := \overline{f\left(\frac{1}{\overline{z}}\right)}, \qquad z \in G^{\#}. \tag{2.2}$$

If the function f is holomorphic (resp. meromorphic) in G, then the 'symmetric' function $f^{\#}$ is holomorphic (resp. meromorphic) in $G^{\#}$. Now we introduce some distinguished classes of functions. We will be concerned with functions which are meromorphic (in particular, holomorphic) in one of the following three domains: the interior $\mathbb{D}_+$ of the unit disc, the exterior $\mathbb{D}_-$ of the unit disc or the (disconnected) set $\mathbb{D}_+ \cup \mathbb{D}_-(= \overline{\mathbb{C}} \backslash \mathbb{T})$.

All the classes considered below will be subclasses of the Nevanlinna class of meromorphic functions of bounded type in the corresponding open set.

Let us give some definitions.

The class $NM(\mathbb{D}_+)$ consists of all functions f which are meromorphic in $\mathbb{D}_+$ and satisfy the following two conditions

$$\overline{\lim}_{r\uparrow 1} \int_{\mathbf{T}} \ell n^+ |f(rt)| m(dt) < \infty, \tag{2.3}$$

$$\sum_k [|1 - |\zeta_k(f)||] < \infty \tag{2.4}$$

where the sum is taken over the set $\{\zeta_k(f)\}$ of poles of the function f (counting multiplicities).

In a symmetric way, the class $NM(\mathbb{D}_-)$ consists of all functions f which are meromorphic in $\mathbb{D}_-$ and fulfill the conditions

$$\overline{\lim}_{r\downarrow 1} \int_{\mathbf{T}} \ell n^+ |f(rt)| m(dt) < \infty, \tag{2.5}$$

$$\sum_k [|1 - |\zeta_k(f)|^{-1}|] < \infty \tag{2.6}$$

where the sum is taken over the set $\{\zeta_k(f)\}$ of poles of the function f (counting multiplicities).

The class $N(\mathbb{D}_+)$ is the subclass of all functions $f \in NM(\mathbb{D}_+)$ holomorphic in $\mathbb{D}_+$. In other words, the class $N(\mathbb{D}_+)$ consists of all functions f which are holomorphic in $\mathbb{D}_+$ and satisfy condition (2.3), which for holomorphic functions f is equivalent to the condition

$$\sup_{0 \leq r < 1} \int_{\mathbf{T}} \ell n^+ |f(rt)| m(dt) < \infty. \tag{2.7}$$

Similarly, the class $N(\mathbb{D}_-)$ is the subclass of all functions $f \in NM(\mathbb{D}_-)$ which are holomorphic in $\mathbb{D}_-$. Thus, the class $N(\mathbb{D}_-)$ consists of all functions f which are holomorphic in $\mathbb{D}_-$ and fulfill condition (2.5), which for holomorphic functions f is equivalent to

$$\sup_{1 < r \leq \infty} \int_{\mathbf{T}} \ell n^+ |f(rt)| m(dt) < \infty. \tag{2.8}$$

Obviously,

$$f \in NM(\mathbb{D}_+) \qquad \Longleftrightarrow \qquad f^\# \in NM(\mathbb{D}_-)$$

and

$$f \in N(\mathbb{D}_+) \qquad \Longleftrightarrow \qquad f^\# \in N(\mathbb{D}_-)$$

The class $N(\mathbb{D}_+)$ of all function of bounded type which are holomorphic in $\mathbb{D}_+$ was introduced in the paper [Ne-Ne] by brothers F. and R. NEVANLINNA, whereas the class $NM(\mathbb{D}_+)$ of all functions of bounded type which are meromorphic in $\mathbb{D}_+$ was considered first by R. NEVANLINNA in [Ne]. In this paper, we will consider functions f which are meromorphic in the disconnected open set $\mathbb{D}_+ \cup \mathbb{D}_-(= \overline{\mathbb{C}}\backslash\mathbb{T})$ and which are *pseudocontinuable*. The property of *pseudocontinuability* is formulated in terms of boundary values.

It is well known that each function f belonging to the meromorphic Nevanlinna class $NM(\mathbb{D}_+)$ has boundary values m-a.e. on $\mathbb{T}$, i.e. for m-almost all $t \in \mathbb{T}$ there exists the radial limit

$$f_+(t) := \lim_{r\uparrow 1} f(rt). \tag{2.9}$$

Similarly, if f belongs to the meromorphic Nevanlinna class $NM(\mathbb{D}_-)$, then, for m-almost all $t \in \mathbb{T}$ there exists the radial limit

$$f_-(t) := \lim_{r\downarrow 1} f(rt). \tag{2.10}$$

Definition 2.1. A function f which is meromorphic in the (disconnected) open set $\mathbb{D}_+ \cup \mathbb{D}_-(= \overline{\mathbb{C}}\backslash\mathbb{T})$ is called *pseudocontinuable*, if $Rstr._{\mathbb{D}_+} f$ and $Rstr._{\mathbb{D}_-} f$ belong to the classes $NM(\mathbb{D}_+)$ and $NM(\mathbb{D}_-)$, respectively, and, moreover, the equality

$$f_+(t) = f_-(t) \tag{2.11}$$

holds for m-almost all $t \in \mathbb{T}$. Here f_+ and f_- are the radial limits of f on $\mathbb{T}$ from the interior and from the exterior of the unit circle, respectively.

Definition 2.2. $PCNM$ is the class of all functions which are meromorphic in $\mathbb{D}_+ \cup \mathbb{D}_-$ and pseudocontinuable.

Definition 2.3. PCN is the subclass of all those functions belonging to $PCNM$ which are holomorphic in $\mathbb{D}_+ \cup \mathbb{D}_-$.

Remark 2.1. A function f belonging to $PCNM$ is originally defined in $\mathbb{D}_+ \cup \mathbb{D}_-$ but, in view of Definition 2.1, for such a function the interior and exterior radial limits $f_+(t)$ and $f_-(t)$ coincide m-a.e. For this reason, a function $f \in PCNM$ can be extended to the points $t \in \mathbf{T}$ where the radial limits $f_+(t)$ and $f_-(t)$ defined by (2.9) and (2.10) exist and satisfy (2.11). For such a point $t \in \mathbf{T}$, we define

$$f(t) := f_+(t)(= f_-(t)) \tag{2.12}$$

If we extend a function $f \in PCNM$ in this way, the extended function will be defined everywhere in the extended complex plane $\overline{\mathbb{C}}$ with the exception of some subset of the unit circle having linear Lebesgue-Borel measure zero. (In particular, this subset can also be empty.)

Remark 2.2. If $f \in PCNM$ is such that for all $z \in \mathbb{D}_+$ we have $f(z) = 0$, then, obviously, $f_+(t) = 0$ for all $t \in \mathbf{T}$. Then because of (2.11) $f_-(t) = 0$ for m-almost all $t \in \mathbf{T}$. Therefore, since f is uniquely determined by its boundary values, $f(z) = 0$ for all $z \in \mathbb{D}_-$. Consequently, a pseudocontinuable function is completely determined by its restriction to the connected component $\mathbb{D}_+$ of its domain.

The class of pseudocontinuable functions was introduced by H.S. Shapiro (see [Sh1], especially item 2.22, p.332 there). In an implicit form, pseudocontinuable functions figured in G.Ts. Tumarkin's paper [Tum2]. However, the class of pseudocontinuable functions was first distinguished explicitly in the paper [Sh1] in 1965. The term "pseudocontinuation" was introduced in this paper. In G.Ts. Tumarkin's papers [Tum2] and [Tum7], pseudocontinuable functions appear in connection with the problem of describing classes of functions which can be approximated by rational functions with prescribed poles. Subsequently this class occured in the papers [DSS1], [DSS2] by R.G. Douglas, H.S. Shapiro and A.L. Shields (as well in the paper [Kr] by T.L. Kriete III) on description of cyclic vectors and invariant subspaces of the backward shift operator, in the paper [RoRo1] by M. Rosenblum and J. Rovnyak on factorization of operator-valued functions which are nonnegative on the unit circle and, finally, in the papers [Ar1], [Ar2] of D.Z. Arov and [DH] of R.G. Douglas and J.W. Helton on Darlington synthesis. (The papers [AFK1, AFK2, AFK3] are also related with Darlington synthesis. Pseudocontinuable functions are of importance there.) See also papers [Mar], [Fuhr], [Lev], [Hi-Wa], [RoRo2], [RoRo3]. Interesting applications of pseudocontinuable functions to the spectral theory of Jakobi matrices are found in [So-Yu].

3. The Smirnov Class N_*.

Definition 3.1. A function $f : \mathbb{D}_+ \to \mathbb{C}$ belongs to the *Smirnov class* $N_*(\mathbb{D}_+)$ if f is holomorphic in $\mathbb{D}_+$ and if the family $\{\ell n^+ |f(rt)|\}_{0 \le r < 1}$ is the uniformly integrable on $\mathbf{T}$ with respect to the normalized Lebesgue-Borel measure m. A function $f : \mathbf{D}_- \to \mathbf{C}$ belongs to the *Smirnov class* $N_*(\mathbf{D}_-)$ if f is holomorphic in $\mathbf{D}_-$ and if the family $\{\ln^+ |f(rt)|\}_{1 < r \leq \infty}$ is uniformly integrable on $\mathbf{T}$ with respect to m.

Uniform integrability means the following: For each $\epsilon > 0$ there exists a $\delta = \delta(\epsilon) > 0$ such that for every Borel subset e of the unit circle $\mathbf{T}$ which satisfies

$$m(e) < \delta \tag{3.1}$$

and for all values of the parameter r indexing the family the inequality

$$\int_e \ln^+ |f(rt)| m(dt) < \epsilon \tag{3.2}$$

is satisfied. (In particular, $\delta(\epsilon)$ is independent on r.)

We have not assumed a priori in definition 3.1 that the function f must belong to the Nevanlinna class $N(\mathbf{D}_+)$ (resp. $N(\mathbf{D}_-)$), but Definition 3.1 as it is given easily implies that
$$N_*(\mathbf{D}_+) \subseteq N(\mathbf{D}_+), \qquad N_*(\mathbf{D}_-) \subseteq N(\mathbf{D}_-).$$
We will deal with the meromorphic classes $NM_*(\mathbf{D}_+)$ and $NM_*(\mathbf{D}_-)$ as well as with the holomorphic Smirnov classes $N_*(\mathbf{D}_+)$ and $N_*(\mathbf{D}_-)$.

Definition 3.2.   A meromorphic function f in $\mathbf{D}_+$ (resp. $\mathbf{D}_-$) is said to belong to the *meromorphic Smirnov class* $NM_*(\mathbf{D}_+)$ (resp. $NM_*(\mathbf{D}_-)$) if the following two conditions are satisfied:

(i) f belongs to $NM(\mathbf{D}_+)$ (resp. $NM(\mathbf{D}_-)$).

(ii) The family $\{ln^+ |f(rt)|\}_r$ is uniformly integrable with respect to the Lebesgue measure on $\mathbf{T}$.

If we speak about $NM_*(\mathbf{D}_+)$, we take $r \in [\frac{1}{2}, 1)$, whereas in the case $NM_*(\mathbf{D}_-)$ the parameter r runs over $(1,2]$.

Obviously,
$$f \in N_*(\mathbf{D}_+) \qquad \Longleftrightarrow \qquad f^\# \in N_*(\mathbf{D}_-)$$
and
$$f \in NM_*(\mathbf{D}_+) \qquad \Longleftrightarrow \qquad f^\# \in NM_*(\mathbf{D}_-).$$

The Smirnov classes are additive and multiplicative, i.e. if $f_1, f_2 \in N_*(\mathbb{D}_+)$, then $f_1 + f_2 \in N_*(\mathbb{D}_+)$, $f_1 \cdot f_2 \in N_*(\mathbb{D}_+)$.

      The fact whether a function f belongs to some distinguished subclass of NM can be characterized with the aid of the Riesz-Nevanlinna-Smirnov factorization. Assume that f belongs to $NM(\mathbb{D}_+)$ and that $f \not\equiv 0$. Then it is well known that f can be represented in the form

$$f(z) = C \cdot \frac{B_1(z)}{B_2(z)} \cdot \exp\left\{-\int_{\mathbb{T}} \frac{t+z}{t-z} \delta_s(dt)\right\} \cdot \exp\left\{\int_{\mathbb{T}} \frac{t+z}{t-z} \ln|f(t)| m(dt)\right\} \quad (z \in \mathbb{D}_+), \quad (3.3)$$

where C is some unimodular constant, B_1 and B_2 are Blaschke products which are built from the zeros and poles of f in $\mathbb{D}_+$ respectively, and δ_s is some signed Borel measure on $\mathbb{T}$ which is singular with respect to the Lebesgue-Borel measure m. (If a function f belongs to $NM(\mathbb{D}_+)$, then the set of its zeros and the set of its poles automatic satisfy the Blaschke condition. Therefore the Blaschke products B_1 and B_2 are well defined and the function $|\ln|f(t)||$ is integrable on $\mathbb{T}$ with respect to m.) Obviously, the Blaschke products B_1 and B_2 and also the singular measure δ_s are determined by f. Of course one or both of the Blaschke, and/or the singular measure can be missing in (3.3). Let f be a function belonging to $NM(\mathbb{D}_+)$ with multiplicative representation (3.3). Clearly,

$$f \in N(\mathbb{D}_+) \qquad \Longleftrightarrow \qquad B_2 \equiv 1. \qquad (3.4)$$

The following assertions are a little bit less obvious, but are nevertheless well known:

$$f \in NM_*(\mathbb{D}_+) \qquad \Longleftrightarrow \qquad \delta_s \geq 0 \qquad (3.5)$$

and

$$f \in N_*(\mathbb{D}_+) \qquad \Longleftrightarrow \qquad (B_2 \equiv 1 \text{ and } \delta_s \geq 0). \qquad (3.6)$$

Suppose $p \in (0, \infty)$.

Definition 3.3. A function $f : \mathbb{D}_+ \to \mathbb{C}$ is said to belong to the *Hardy class* $H^p(\mathbb{D}_+), 0 < p < \infty$, if f is holomorphic in $\mathbb{D}_+$ and if

$$\sup_{r \in [0,1)} \int_{\mathbb{T}} |f(rt)|^p m(dt) < \infty. \qquad (3.7)$$

We say that a function $f : \mathbb{D}_- \to \mathbb{C}$ belongs to the *Hardy class* $H^p(\mathbb{D}_-), 0 < p < \infty$, if f is holomorphic in $\mathbb{D}_-$ and if

$$\sup_{r \in (1,\infty]} \int_{\mathbb{T}} |f(rt)|^p m(dt) < \infty. \qquad (3.8)$$

A function $f : \mathbf{D}_+ \to \mathbf{C}$ is referred to as belonging to the *Hardy class* $H^\infty(\mathbf{D}_+)$ if f is holomorphic and bounded in $\mathbf{D}_+$, i.e. if

$$\sup_{z \in \mathbf{D}_+} |f(z)| < \infty. \tag{3.9}$$

Finally, a function $f : \mathbf{D}_- \to \mathbf{C}$ is referred to as belonging to the *Hardy class* $H^\infty(\mathbf{D}_-)$ if f is holomorphic and bounded in $\mathbf{D}_-$, i.e. if

$$\sup_{z \in \mathbf{D}_-} |f(z)| < \infty. \tag{3.10}$$

Obviously, $f \in H^p(\mathbf{D}_+)$ $\Longleftrightarrow$ $f^\# \in H^p(\mathbf{D}_-)$.

The following fact which is due to V.I. Smirnov is of principal importance for us.

Maximum principle of V.I. Smirnov: *Suppose that the function f belongs to $N_*(\mathbf{D}_+)$ and that for its boundary values $f(t) := \lim_{r \uparrow 1} f(rt)$ and for some $p \in (0, \infty)$ the condition*

$$\int_{\mathbf{T}} |f(t)|^p m(dt) < \infty \tag{3.11}$$

is satisfied. Then $f \in H^p(\mathbf{D}_+)$. If

$$\operatorname*{ess\,sup}_{t \in \mathbf{T}} |f(t)| < \infty, \tag{3.12}$$

then $f \in H^\infty(\mathbf{D}_+)$.

The class $N_*(\mathbf{D})$ was introduced by V.I. SMIRNOV in [Sm]. This paper also contains the theorem which we have called "maximum principle of V.I. Smirnov". (Smirnov himself chose the symbol D for the class which we have denoted by N_*. For this class one can also often find the symbols N_+ or N^+.) To our knowledge, the meromorphic Smirnov class (which we have denoted by NM_*) was not considered before. For the basic facts on the classes of Nevanlinna, Smirnov and Hardy we refer the reader to the monographs of P.L. Duren [Dur], J.B. Garnett [Gar] and P. Koosis [Koo2]. In particular, the maximum principle of V.I. Smirnov is presented as Theorem 2.11 in [Dur]. A selection of papers by V.I. Smirnov on complex analysis, including comments on the further progress, is contained in the article of N.K. Nikol'skiĭ and V.P. Khavin [Ni-Kh]. The monograph by W. Rudin [Rud2] deals with classes of functions of several variables in the polydisc which can be conceived as natural analogues of the classes of Nevanlinna, Smirnov and Hardy. The results presented there are also meaningful for functions of one variable. In [Rud2], there is also a well-written presentation of special questions concerning functions of one variable (see Chapter III). Several aspects of the theory of Hardy spaces are also contained in the monograph by K. Hoffman [Hoff].

4. The weighted space $PCH_w^p(I^+, I^-)$ of pseudocontinuable meromorphic functions with prescribed denominators.

Definition 4.1. Let I^+ [resp. I^-] be a given inner function in $\mathbb{D}_+$ [resp. $\mathbb{D}_-$]. This pair (I^+, I^-) is called *a pair of denominators*. I^+ [resp. I^-] is called a *plus-denominator* [resp. *minus-denominator*]. Let f be a function belonging to the class $PCNM$ defined in item 2. (i.e. f is defined everywhere in the extended complex plane $\overline{\mathbb{C}}$ with the exeption of some subset of $\mathbf{T}$ having linear Lebesgue-Borel measure zero: see Remark 2.1). Let

$$f_+ := Rstr._{\mathbb{D}_+} f, \qquad f_- := Rstr._{\mathbb{D}_-} f \qquad (4.1)$$

By definition, *the function f belongs to the pair (I^+, I^-) of denominators*, if

$$f_+ I^+ \in N_*(\mathbb{D}_+), \qquad f_- I^- \in N_*(\mathbb{D}_-) \qquad (4.2)$$

The class of all pseudocontinuable meromorphic functions belonging to given pair (I^+, I^-) of denominators is designated by $PCNM(I^+, I^-)$

Obviously $PCNM(I^+, I^-) \subseteq PCNM$.

Remark 4.1. The function

$$n_+(z; f) := f_+(z) I^+(z) \qquad (z \in \mathbb{D}_+) \qquad (4.3^+)$$

belonging to the Smirnov class $N_*(\mathbb{D}_+)$ (see (4.2)) admits the inner-outer factorization

$$n_+(z; f) = I^+(z; f) E^+(z, f) \qquad (z \in \mathbb{D}_+). \qquad (4.4^+)$$

Thus the function f_+, which is meromorphic in $\mathbb{D}$, is representable in the form

$$f_+(z)) = \frac{I^+(z; f)}{I^+(z)} E^+(z; f) \qquad (z \in \mathbb{D}_+). \qquad (4.5^+)$$

where $I^+(.)$ is the given inner function - the first component of the pair (I^+, I^-), $I^+(.; f)$ is an inner function, and $E^+(.; f)$ is an outer one in $\mathbb{D}_+$. We do not assume that the inner function $I^+(.; f)$ and $I^+(.)$ are relatively prime: The functions $I^+(.; f)$ and $I^+(.)$

can have a nonconstant common divisor. However, even if the ratio $I^+(.; f)/I^+$ is reducible, we do not reduce it and prefer to have the denominator in (4.5^+) equal to the first component of the given pair (I^+, I^-). Similarly, the function

$$n_-(z; f) = f_-(z)I^-(z) \qquad (z \in \mathbb{D}_-), \qquad (4.3^-)$$

belongs to the Smirnov class (see (4.2)), and admits the inner-outer factorization

$$n_-(z; f) = I^-(z; f)E^-(z; f) \qquad (z \in \mathbb{D}_-). \qquad (4.4^-)$$

Thus

$$f_-(z) = \frac{I^-(z; f)}{I^-(z)}E^-(z; f) \qquad (z \in \mathbb{D}_-), \qquad (4.5^-)$$

where I^- is second component of the given pair (I^+, I^-). Here we also shall keep this representation even if $I^-(z; f)$ and $I^-(f)$ have common divisors.

Definition 4.2. Let (I^+, I^-) be a given pair of denominators (i.e. $I^\pm$ are inner in $\mathbb{D}^\pm$). Let $w \in \mathcal{W}$ be a weight function, and $p \in (0, \infty)$. Let f be a pseudocontinuable meromorphic function, i.e., $f_+ \in NM(\mathbb{D}_+), f_- \in NM(\mathbb{D}_-)$ (f_+ and f_- as in (4.1), and (2.11) is fulfilled). Then *the function f belongs to the class* $PCH_w^p(I^+, I^-)$ if the following additional conditions are satisfied:

i). $f \in PCNM(I^+, I^-)$, i.e. the condition (4.2) is fulfilled;

ii).   *Rstr.*$_\mathbb{T}f$ belongs to the weighted space L_w^p, i.e.

$$\int_{\mathbb{T}} |f(t)|^p w(t)m(dt) < \infty \qquad (4.6)$$

(we recall that the function $f \in PCNM$ is defined a.e. on $\mathbb{T}$).

In other words,

$$PCH_w^p(I^+, I^-) := PCNM(I^+, I^-) \cap L_w^p \qquad (4.7)$$

Particular cases of $PCH_w^2(I^+, I^-)$ have been considered earlier: for the special weight function by Kriete III [Kr], and for general weight function $w \in \mathcal{W}$, but with I^+ and I^- restricted to be convergent Blashke products, by Katsnelson [Kats2] (See item 5 there. There the space was referred to as $PCH_w^2(S)$).

Obviously, the space $PCH_w^p(I^+, I^-)$ equipped with the natural linear operations is a complex vector space. We equip it with the metric induced from the corresponding space L_w^p. If $f, g \in PCH_w^p(I^+, I^-)$, then the distance between them is defined by usual rules:

$$\rho_{L_w^p}(f, g) := \begin{cases} \int_{\mathbb{T}} |f(t) - g(t)|^p w(t)(dt), & \text{if } p \in (0, 1) \\ \{\int_{\mathbb{T}} |f(t) - g(t)|^p w(t) m(dt)\}^{\frac{1}{p}}, & \text{if } p \in [1, \infty). \end{cases} \tag{4.8}$$

The logarithmic integral converges for each nonzero function in the Nevanlinna class NM. Therefore, the same is true in the class $PCH_w^p(I^+, I^-)$:

$$\int_{\mathbb{T}} |\ell n |f(t)|| m(dt) < \infty \quad \text{for all } f \in PCH_w^p(I^+, I^-), \quad f \not\equiv 0. \tag{4.9}$$

Thus, the convergence of the logarithmic integral

$$\int_{\mathbb{T}} |\ell n w(t)| m(dt) < \infty \tag{4.10}$$

is a necessary condition for the space $PCH_w^p(I^+, I^-)$ to be nontrivial (i.e. to contain a nonzero element). The convergence of the logarithmic integral implies that $w(t) > 0$ a.e. on $\mathbb{T}$. Therefore, if $\rho(f, g)_{L_w^p} = 0$, then $f(t) = g(t)$ a.e. on $\mathbb{T}$. From the boundary uniqueness theorem we obtain that $f \equiv g$. Hence, elements of the space $PCH_w^p(I^+, I^-)$ can be considered as *functions* defined a.e. on $\mathbb{T}$ (not only as *classes* consisting of equivalent functions).

Thus, *the space $PCH_w^p(I^+, I^-)$ equipped with the natural linear operations and with the metric (4.8) is a metric vector space.*

Theorem 4.1. *The metric space $PCH_w^p(I^+, I^-)$ is separable and complete.*

Proof. As the space L_w^p is separable and $PCH_w^p(I^+, I^-)$ is a subspace of L_w^p, it is separable. To prove completeness, suppose that $\{f_n\}_{n \in \mathbb{N}}$ is a Cauchy sequence belonging to $PCH_w^p(I^+, I^-)$, i.e.

$$\int_{\mathbb{T}} |f_n(t) - f_l(t)|^p w(t) m(t) \to 0 \quad (n, l \to \infty). \tag{4.11}$$

By a theorem of G. Szegö [Sz] (see also [Dur, item 2.4], [Gar, Ch.2, section 4], [Koo, Ch.IV, D]), a weight function satisfying (4.10) admits a "factorization"

$$w(t) = |\Phi^+(t)|^p, \qquad w(t) = |\Phi^-(t)|^p \quad \text{(for a. e. } t \in \mathbb{T}) \tag{4.12}$$

in terms of boundary values $\Phi^+(t), \Phi^-(t)$ of

$$\Phi^+(z) := \exp\{\frac{1}{p} \int_{\mathbb{T}} \frac{t + z}{t - z} [\ell n w(t)] m(dt)\}, \qquad (z \in \mathbb{D}_+)$$

and   (4.13)

$$\Phi^-(z) := \exp\{\frac{1}{p}\int_{\mathbb{T}}\frac{z+t}{z-t}[\ell nw(t)]m(dt)\}, \qquad (z \in \mathbb{D}_-)$$

respectively. The functions Φ^+ and Φ^- are outer, i.e. satisfy the conditions

$$\Phi^+ \in N_*(\mathbb{D}_+), \quad \frac{1}{\Phi^+} \in N_*(\mathbb{D}_+), \quad \Phi^- \in N_*(\mathbb{D}_-), \quad \frac{1}{\Phi^-} \in N_*(\mathbb{D}_-) \qquad (4.14)$$

and $\Phi^- = (\Phi^+)^\#$.

For $n \in \mathbb{N}$ we set

$$\left.\begin{array}{ll} h_{n,+} := f_{n,+}(z)I^+(z)\Phi^+(z), & (z \in \mathbb{D}_+), \\ h_{n,-} := f_{n,-}(z)I^-(z)\Phi^-(z), & (z \in \mathbb{D}_-), \end{array}\right\} \qquad (4.15)$$

where $f_{n,+} := Rstr._{\mathbb{D}_+} f_n, \quad f_{n,-} := Rstr._{\mathbb{D}_-} f_n.$ As $f_n \in PCH_w^p,$

$$f_{n,+}I^+ \in N_*(\mathbb{D}_+), \qquad f_{n,-}I^- \in N_*(\mathbb{D}_-).$$

Then, since $\Phi^+ \in N_*(\mathbb{D}_+), \ \Phi^- \in N_*(\mathbb{D}_-)$, we see that

$$h_{n,+} \in N_*(\mathbb{D}_+), \qquad h_{n,-} \in N_*(\mathbb{D}_-) \qquad (4.16)$$

and in view of (4.12)

$$|h_{n,+}(t)|^p = |f_n(t)|^p w(t) \qquad (a.e.)$$

$$|h_{n,-}(t)|^p = |f_n(t)|^p w(t) \qquad (a.e.)$$

Since $f_n \in L_w^p$, we have

$$\int_{\mathbb{T}}|h_{n,\pm}|^p m(dt) = \int_{\mathbb{T}}|f_n(t)|^p w(t)m(dt) < \infty$$

In the other words,

$$h_{n,+} \in L_w^p, \qquad h_{n,-} \in L_w^p \qquad (4.17)$$

Therefore, from (4.16), (4.17) and Maximum principle of V.I. Smirnov it follows that

$$h_{n,+} \in H^p(\mathbb{D}_+), \qquad h_{n,-} \in H^p(\mathbb{D}_-) \qquad (4.18)$$

Thus condition (4.11) splits into the conditions

$$\int_{\mathbb{T}}|h_{n,+}(t) - h_{l,+}(t)|^p m(dt) \to 0 \qquad (n,l \to \infty)$$

$$\int_{\mathbb{T}} |h_{n,-}(t) - h_{l,-}(t)|^p m(dt) \to 0 \qquad (n, l \to \infty),$$

which implies that $\{h_{n,+}\}_n$ is a Cauchy sequence in $H^p(\mathbb{D}_+)$, and $\{h_{n,-}\}_n$ is a Cauchy sequence in $H^p(\mathbb{D}_-)$. Since the space $H^p(\mathbb{D}_+)$ and $H^p(\mathbb{D}_-)$ are complete metric spaces, there exist limit functions

$$h_+ \in H^p(\mathbb{D}_+), \qquad h_- \in H^p(\mathbb{D}_-) \tag{4.19}$$

such that

$$\left. \begin{aligned} \int_{\mathbb{T}} |h_{n,+}(t) - h_+(t)|^p m(dt) \to 0 \qquad (n \to \infty), \\[2mm] \int_{\mathbb{T}} |h_{n,-}(t) - h_-(t)|^p m(dt) \to 0 \qquad (n \to \infty). \end{aligned} \right\} \tag{4.20}$$

We set

$$\left. \begin{aligned} f_+(z) &:= h_+(z)(\Phi^+(z))^{-1}(I^+(z))^{-1} \qquad (z \in \mathbb{D}_+), \\ f_-(z) &:= h_-(z)(\Phi^-(z))^{-1}(I^-(z))^{-1} \qquad (z \in \mathbb{D}_-) \end{aligned} \right\} \tag{4.21}$$

and

$$f(z) := \begin{cases} f_+(z), & z \in \mathbb{D}_+ \\ f_-(z), & z \in \mathbb{D}_- \end{cases} . \tag{4.22}$$

In view of (4.19) and (4.14),

$$h_+(\Phi^+)^{-1} \in N_*(\mathbb{D}_+), \qquad h_-(\Phi^-)^{-1} \in N_*(\mathbb{D}_-) \tag{4.23}$$

Equalities (4.21), (4.23) mean that

$$f_+ I^+ \in N_*(\mathbb{D}_+), \qquad f_- I^- \in N_*(\mathbb{D}_-) \tag{4.24}$$

From (4.12), (4.15), (4.21) and (4.20) it follows that

$$\int_{\mathbb{T}} |f_{n,+}(t) - f_+(t)|^p w(t) m(dt) \to 0 \qquad (n \to \infty)$$

$$\int_{\mathbb{T}} |f_{n,-}(t) - f_-(t)|^p w(t) m(dt) \to 0 \qquad (n \to \infty)$$

As $f_{n,+}(t) = f_{n,-}(t) = f_n(t)$ (it is condition (2.11) for the function f_n), we have

$$\left. \begin{aligned} \int_{\mathbb{T}} |f_n(t) - f_+(t)|^p w(t) m(dt) \to 0 \qquad (n \to \infty) \\[2mm] \int_{\mathbb{T}} |f_n(t) - f_-(t)|^p w(t) m(dt) \to 0 \qquad (n \to \infty) \end{aligned} \right\} \tag{4.25}$$

From (4.25) it follows that

$$\int_{\mathbb{T}} |f_+(t) - f_-(t)|^p w(dt) = 0$$

As $w(t) > 0$ a.e.,

$$f_+(t) = f_-(t) \qquad \text{(a.e.)} \qquad\qquad (4.26)$$

Conditions (4.26) and (4.24) imply that $f \in PCNM(I^+, I^-)$, whereas conditions (4.19), (4.21), and (4.12) imply that $f \in L_w^p$. Thus, $f \in PCH_w^p(I^+, I^-)$. Condition (4.25) means that $\int_{\mathbb{T}} |f_n(t) - f(t)|^p w(t) m(dt) \to 0 \quad (n \to \infty)$. By (4.8), $\rho_{L_w^p}(f_n, f) \to 0 \quad (n \to \infty)$. The theorem is proved. $\square$

Example 4.1. i) Let $I^\pm$ be given inner functions in $\mathbb{D}_\pm$. Then in fact, each of these functions is defined and meromorphic in $\mathbb{D}_+ \cup \mathbb{D}_-(= \overline{\mathbb{C}} \backslash \mathbb{T})$, and is pseudocontinuable: $I^\pm = 1/\overline{I^\pm(1/\overline{z})}$, $(z \in \mathbb{D}_+ \cup \mathbb{D}_-)$. Set

$$f(z) = \frac{1}{2}((I^+(z))^{-1} + (I^-(z))^{-1}) \qquad (z \in \overline{\mathbb{C}} \backslash \mathbb{T}). \qquad (4.27)$$

The function f is pseudocontinuable and meromorphic in $\mathbb{D}_+ \cup \mathbb{D}_-$. Moreover, it belongs to $PCNM(I^+, I^-)$. Indeed

$$I^+(z)f_+(z) = \frac{1}{2}(1 + I^+(z)\overline{I^-(1/\overline{z})}) \qquad (z \in \mathbb{D}_+)$$

and hence $|I^+(z)f_+(z)| \le 1 \quad (z \in \mathbb{D}_+)$, $I^+ f_+ \in N_*(\mathbb{D}_+)$. Similarly, $I^- f_- \in N_*(\mathbb{D}_-)$. As $|f(t)| \le 1 \quad (t \in \mathbb{T})$, the function $f(t)$ belongs to every weighted space L_w^p with a summable weight function. If at least one of the functions $I^\pm$ is not constant, then the function f defined by (4.27) is not constant. In this case (if also (1.4) is in force) the space $PCH_w^p(I^+, I^-)$ contains a nonconstant function.

ii). If f is a function from $PCH_w^p(I^+, I^-)$, $z_0 \in \mathbb{D}_+ \cup \mathbb{D}_-$, and z_0 is not a pole of the function f, then the function $\frac{f(z)-f(z_0)}{z-z_0}$ belongs to the space $PCH_w^p(I^+, I^-)$ as well. Thus, if the space $PCH_w^p(I^+, I^-)$ contains a nonconstant function, then the dimension of this space is greater than one.

Remark 4.2. If the two functions I^+ and I^- are constant: $I^+ \equiv 1$ and $I^- \equiv 1$, then $f_+ \in N_*(\mathbb{D}_+)$ and $f_- \in N_*(\mathbb{D}_-)$ for each $f \in PCNM(I^+, I^-)$. If $p > 1$ and the weight function w satisfies the condition

$$\int_{\mathbb{T}} [w(t)]^{-\frac{1}{p-1}} m(dt) < \infty \qquad (4.28)$$

then $\int_{\mathbb{T}} |f(t)|m(dt) < \infty$ for every function $f \in L_w^p$. Thus, if $I^+ \equiv 1$, $I^- \equiv 1$, and the weight function w satisfies condition (4.28), then $f_+ \in N_*(\mathbb{D}_+)$, $f_- \in N_*(\mathbb{D}_-)$ and $\int_{\mathbb{T}} |f(t)|m(dt) < \infty$. By the Maximum Principle of V.I. Smirnov, $f_+ \in H^1(\mathbb{D}_+)$, $f_- \in H^1(\mathbb{D}_-)$ for each $f \in PCH_w^p(I^+, I^-)$. The boundary values of these functions f_+ and f_- coincide a.e. on $\mathbb{T}$. Consequently, a variant of the Painleve Theorem on removing singularities implies that these functions can be analytically continued through $\mathbb{T}$ into each other (see [Koo2, Theorem III.E.2]). For this reason, the original function f is holomorphic in the extended complex plane and must be constant.

5. G. Ts. Tumarkin's theorem on functions which admit weighted approximation by a sequence of rational functions with preassigned poles.

Now we are able to formulate G.Ts. Tumarkin's theorem.

Theorem 5.1. *Let $\mathcal{J}$ be a table $[a_{kj}]$ of complex numbers of the form (1.1) which is subject to the restrictions (1.2) and (1.7). Let f be a given function from the weighted space L_w^p, where $p \in (0, \infty)$ and $w \in \mathcal{W}$. Suppose that there exists a sequence $\{R_k\}_{k \in \mathbb{N}}$ of rational function in $\mathcal{R}(\mathcal{J})$ such that*

$$\lim_{k \to \infty} \int_{\mathbb{T}} |f(t) - R_k(t)|^p w(t)m(dt) = 0 \qquad (5.1)$$

Then the function f belongs to the space $PCH_w^p(I_{\mathcal{J}}^+, I_{\mathcal{J}}^-)$, where $I_{\mathcal{J}}^+$ and $I_{\mathcal{J}}^-$ are the plus-denominator and the minus-denominator corresponding to the table $\mathcal{J}$. (This means that $f(t) = \lim_{r \uparrow 1} f_+(rt) = \lim_{r \downarrow 1} f_-(rt)$ a.e. on $\mathbb{T}$, where f is a pseudocontinuable meromorphic function belonging to the class $PCNM(I_{\mathcal{J}}^+, I_{\mathcal{J}}^-)$.)

Theorem 5.1 is formulated and proved in the paper [Tum7] for $p \in [1, \infty)$ only. The proof uses the duality between L^p and L^q, $(\frac{1}{p} + \frac{1}{q} = 1)$. The proof for general $p \in (0, \infty)$ was given in [Tum9], and has not been published elsewhere. The proof in [Tum9] is based on Tumarkin's results on convergent sequence of meromorphic functions.

Remark 5.1. Theorem 5.1 claims in particular that if a sequence $\{R_k\}_{k \in \mathbb{N}}$ with preassigned poles satisfying condition (1.7) converges (with respect to L_w^p-topology) on the unit circle, then this sequence converges on $\mathbb{D}_+ \cup \mathbb{D}_-$ too. It is a particular example of the general phenomenon on overconvergence, which was discovered by A. Ostrowskiĭ in [O]. A general point of view on overconvergence was given in the papers [Sh2], [Sh3] by H.S. Shapiro; see also the paper [Gon] by A.A. Gončar.

Theorem 5.1 can also be formulated in the language of spectral function theory. To do this, we first recall (see Nikol'skiĭ's book [Nik2]), that for any inner function Θ^+ in $\mathbf{D}_+$,

$$\mathbf{K}_{\Theta^+} := H^2(\mathbf{D}_+) \ominus \Theta^+ H^2(\mathbf{D}_+) \tag{5.2$^+$}$$

$$\mathbf{K}_{\Theta^-} := H^2(\mathbf{D}_-) \ominus \Theta^- H^2(\mathbf{D}_-) \tag{5.2$^-$}$$

(The orthogonal complement here is with respect to the standard scalar product in the Hardy space $L^2(\mathbf{T})$.) It is well known that every function from the space $\mathbf{K}_{\Theta^+}$ is pseudo-continuable, and that

$$\mathbf{K}_{\Theta^+} = \{f \in L^2 : \ f(t) \in H^2(\mathbf{D}_+),\ t\overline{\Theta}(t)f(t) \in H^2(\mathbf{D}_-)\}. \tag{5.3}$$

(See, for example, [Nik2], Lecture 2, item 1, Lemma on $Lat\ S$.) Thus, if $f \in \mathbf{K}_{\Theta^+}$ then there exist a function $n_- \in H^2(\mathbf{D}_-)$ such that

$$f(t) = t^{-1}\Theta^+(t)n_-(t) \qquad \text{(a. e. on } \mathbf{T}),$$

and hence that

$$r(z) := \begin{cases} f(z), & z \in \mathbf{D}_+ \\[2mm] \dfrac{n_-(z)}{z\Theta^+(1/\overline{z})}, & z \in \mathbf{D}_- \end{cases}$$

is pseudocontinuable: $\lim_{\rho\uparrow 1} r(\rho t) = \lim_{\rho\downarrow 1} r(\rho t)$ $\quad$ (a.e.)

Let $\mathbf{C}$ be the subspace of the space L^p_w which consists of all the constant functions. It is not difficult to see that

$$PCH^2_{\mathbf{1}}(1, (\Theta^+)^\#) = \mathbf{K}_{\Theta^+} + \mathbf{C} \tag{5.4$^+$}$$

($w(t) \equiv 1$ now), and similarly that

$$PCH^2_{\mathbf{1}}((\Theta^-)^\#, 1) = \mathbf{K}_{\Theta^-} + \mathbf{C} \tag{5.4$^-$}$$

The spaces $\mathbf{K}_{\Theta^+}$ and $\mathbf{K}_{\Theta^-}$ are subspaces of the space L^2. We would like to define the spaces $\mathbf{K}_{\Theta^+}$ and $\mathbf{K}_{\Theta^-}$ as subspaces of the weighted space L^p_w ($p \in (0,\infty)$, $w \in W$). To give these definitions in a natural way we associate with the given inner functions Θ^+, Θ^- the kernels

$$\left.\begin{array}{ll} K_{\Theta^+}(t,\zeta) := \dfrac{1 - \Theta^+(t)\overline{\Theta^+(\zeta)}}{1 - t\overline{\zeta}}, & (t \in \mathbf{T}) \\[6mm] K_{\Theta^-}(t,\zeta) := \dfrac{1 - \Theta^-(t)\overline{\Theta^-(\zeta)}}{1 - (t\overline{\zeta})^{-1}}. & (t \in \mathbf{T}) \end{array}\right\} \tag{5.5}$$

From the outset we consider the inner functions $\Theta^{\pm}$ as functions defined on the set $\mathbb{D}_+ \cup \mathbb{D}_-$ and pseudocontinuable:

$$\Theta^+(z) = 1/\overline{\Theta^+(1/\bar{z})}, \quad \Theta^-(z) = 1/\overline{\Theta^-(1/\bar{z})} \qquad (z \in \mathbb{D}_+ \cup \mathbb{D}_-)$$

Thus, the function Θ^+[resp. Θ^-] is holomorphic in $\mathbb{D}_+$ and meromorphic in $\mathbb{D}_-$ [resp. holomorphic in $\mathbb{D}_-$ and meromorphic in $\mathbb{D}_+$]. Hence, the kernel $K_{\Theta\pm}(t,\zeta)$ is defined for all the points $\zeta \in \mathbb{D}_+ \cup \mathbb{D}_-$, which are not poles of of the functions $\Theta^{\pm}$. For every $\zeta \in \mathbb{D}_+ \cup \mathbb{D}_-$ which is not a pole of $\Theta^{\pm}$ the functions $K_{\Theta\pm}$ are bounded on $\mathbb{T}$:

$$\sup_{t \in \mathbb{T}} |K_{\Theta\pm}(t;\zeta)| \leq \frac{1 + |\Theta^{\pm}(\zeta)|}{|1 - |\zeta|^{\pm 1}|}$$

Thus, for every such ζ the kernels $K_{\Theta\pm}(t,\zeta)$ belong to the space L_w^p. Moreover, it is easy to see that $K_{\Theta+}(t,\zeta) \in H^\infty(\mathbb{D}_+)$ and $(\Theta^+(t))^{\#} K_{\Theta+}(t,\zeta) \in H^\infty(\mathbb{D}_-)$. Hence

$$K_{\Theta+}(t,\zeta) \in PCH_w^p(1,(\Theta^+)^{\#}) \qquad (\zeta \in \mathbb{D}_+ \cup \mathbb{D}_-, \ \zeta \text{ is not a pole of } \Theta^+). \tag{5.6$^+$}$$

Similarly,

$$K_{\Theta-}(t,\zeta) \in PCH_w^p((\Theta^-)^{\#},1) \qquad (\zeta \in \mathbb{D}_+ \cup \mathbb{D}_-, \ \zeta \text{ is not a pole of } \Theta^-) \tag{5.6$^-$}$$

It is not difficult to see that the space $\mathbf{K}_{\Theta+}$ (defined by (5.2$^+$) or by (5.3)) is the closed (with respect to L^2-topology) linear span of the set $K_{\Theta+}(t,\zeta)$ (ζ runs over $\mathbb{D}_+ \cup \mathbb{D}_-$, ζ is not a pole of Θ^+):

$$\mathbf{K}_{\Theta+} = \overline{\bigvee_\zeta K_{\Theta+}(t,\zeta)}. \tag{5.7$^+$}$$

Similarly,

$$\mathbf{K}_{\Theta-} = \overline{\bigvee_\zeta K_{\Theta-}(t,\zeta)}. \tag{5.7$^-$}$$

(The function $K_{\Theta\pm}(.,\zeta)$ is a reproducing kernel in the space $\mathbf{K}_{\Theta\pm}$. In fact, it is not necessary to consider all points ζ from $\mathbb{D}_+ \cup \mathbb{D}_-$ in (5.7). It is enough to choose some set of points ζ which does not satisfy the Blaschke condition and does not intersect the set of poles.)

Definition 5.1. *Let $w \in W$, and let $p \in (0,+\infty)$. Let $\Theta^{\pm}$ be an inner function in $\mathbb{D}_{\pm}$. Then the space $\mathbf{K}_w^p(\Theta^{\pm})$ is defined as the closed (with respect to L_w^p-topology) linear span of the set $\{K_{\Theta\pm}(t,\zeta)\}_{\zeta \in \mathbb{D}_+ \cup \mathbb{D}_-}$ of kernels of the form (5.5):*

$$\mathbf{K}_w^p(\Theta^+) := \overline{\bigvee_\zeta K_{\Theta+}(t,\zeta)}, \quad \mathbf{K}_w^p(\Theta^-) := \overline{\bigvee_\zeta K_{\Theta-}(t,\zeta)}. \tag{5.8}$$

Lemma 5.1. *If* $w \in W$, *then the space* $\mathbf{K}_w^p(\Theta^+)$ *is a subspace of the space* $PCH_w^p(1,(\Theta^+)^\#)$, *and the space* $\mathbf{K}_w^p(\Theta^-)$ *is a subspace of the space* $PCH_w^p((\Theta^-)^\#,1)$:

$$\mathbf{K}_w^p(\Theta^+) + \mathbf{C} \subseteq PCH_w^p(1,(\Theta^+)^\#), \quad \mathbf{K}_w^p(\Theta^-) + \mathbf{C} \subseteq PCH_w^p((\Theta^-)^\#,1). \tag{5.9}$$

Proof. The lemma follows immediately from (5.6) and from the completeness of the spaces $PCH_w^p((1,\Theta^+)^\#)$ and $PCH_w^p((\Theta^-)^\#,1)$ □

Definition 5.2. *Let* $w \in W$, $p \in (0,\infty)$ *and let* $\Theta^\pm$ *be an inner functions in* $\mathbf{D}^\pm$. *Then the space* $\mathbf{K}_w^p(\Theta^+,\Theta^-)$ *is defined as the closed (with respect to* L_w^p-*topology) linear span*

$$\mathbf{K}_w^p(\Theta^+,\Theta^-) := \overline{\mathbf{K}_w^p(\Theta^+) \vee \mathbf{K}_w^p(\Theta^-)} \tag{5.10}$$

or (what is the same)

$$\mathbf{K}_w^p(\Theta^+,\Theta^-) = \overline{\vee_{\zeta \in \mathbf{D}_+ \cup \mathbf{D}_-}(K_{\Theta^+}(t,\zeta) \vee K_{\Theta^-}(t,\zeta))} \tag{5.10'}$$

Lemma 5.2. *If* $w \in W$ *and* $p \in (0,\infty)$, *then the space* $\mathbf{K}_w^p(\Theta^+,\Theta^-)$ *is a subspace of the space* $PCH_w^p((\Theta^-)^\#,(\Theta^+)^\#)$:

$$\mathbf{K}_w^p(\Theta^+,\Theta^-) + \mathbf{C} \subseteq PCH_w^p((\Theta^-)^\#,(\Theta^+)^\#) \tag{5.11}$$

Warning. If $p = 2$ and $w \equiv 1$, then it is well known that

$$\mathbf{K}_{\mathbf{1}}^2(\Theta^+) + \mathbf{C} = PCH_{\mathbf{1}}^2(1,(\Theta^+)^\#), \quad \mathbf{K}_{\mathbf{1}}^2(\Theta^-) + \mathbf{C} = PCH_{\mathbf{1}}^2((\Theta^-)^\#,1)$$

(that is the same that (5.4)). However the similar statement is <u>not</u> true for a weighted space (i.e. $w \not\equiv 1$). The inclusions in (5.11) are proper in general.

Lemma 5.3. *Let* $S = \{a_j\}$ *be a finite subset of distinct points in* $\mathbf{D}_+ \cup \mathbf{D}_-$, *let* $S_\pm := S \cap \mathbf{D}_\pm$, *and set*

$$B^\pm(z) = \prod_{a_j \in S_\pm} \frac{a_j - z}{1 - \overline{a_j}z} \cdot \frac{|a_j|}{a_j}. \tag{5.12}$$

(The functions $(B^\pm)^\#(z) = \prod_{a_j \in S_\pm} \frac{1-\overline{a_j}z}{a_j-z} \cdot \frac{|a_j|}{a_j}$ *are finite Blaschke products in* $\mathbf{D}_\mp$.) *Then for each* $p \in (0,\infty)$ *and for every summable weight function* w, *the space* $\mathbf{K}_w^p((B^\pm)^\#)+\mathbf{C}$ *coincides with the set of all rational functions* R *whose poles belong to the set* $S_\pm$:

$$\mathbf{K}_w^p((B^\pm)^\#) + \mathbf{C} = \{R: R(t) = c_0 + \sum_{a_j \in S_\pm} \frac{c_j}{t - a_j}\} \tag{5.13}$$

and

$$\mathbf{K}_w^p((B^-)^\#,(B^+)^\#) + \mathbf{C} = \{R: \ R(t) = c_0 + \sum_{a_j \in S} \frac{c_j}{t - a_j}\} \qquad (5.14)$$

(In (5.13) and (5.14) the poles $\{a_j\}$ are prescribed and the coefficients $\{c_j\}$ remain to be disposed of.)

Lemma 5.3 is well known (see, for example, [Nik2, Lecture II, item 2, Lemma on finite dimensional subspace]); it has a purely algebraic character and does not depend on the choice of the weight function w or the choice of $p \in (0, \infty)$.

Thus, instead of considering a table $\mathcal{T}$ of the form (1.1) (with the constraint (1.2)), we can consider the sequence $\{B_k^+, B_k^-\}_{k \in \mathbb{N}}$ of corresponding finite Blaschke products (1.10) and the sequence of subspaces $\{\mathbf{K}_w^p(B_k^-)^\#, (B_k^+)^\#\}_{k \in \mathbb{N}}$ of space L_w^p. (The space $\mathbf{K}_w^p((B_k^-)^\#, (B_k^+)^\#)$ is the set of all rational functions R_k of the form (1.3).)

More generally, we can consider a given sequence $\{I_k^+, I_k^-\}_{k \in \mathbb{N}}$ of pairs of inner functions in $\mathbb{D}_+$ and $\mathbb{D}_-$ as the primary data. We associate with each pair $\{I_k^+, I_k^-\}$ the subspace $\mathbf{K}_w^p((I_k^-)^\#, (I_k^+)^\#) \subseteq PCH_w^p(I_k^+, I_k^-)$ (see Lemma 5.2). To formulate the generalization of Tumarkin's theorem, we have to define the plus- and minus denominators for a given sequence $\{I_k^+, I_k^-\}_{k \in \mathbb{N}}$. We suppose that

$$\overline{\lim_{k \to \infty}} |I_k^\pm(z)| \not\equiv 0 \qquad (z \in \mathbb{D}_\pm). \qquad (5.15)$$

(The functions $I_k^\pm$ are considered in $\mathbb{D}_\pm$ only.) Now we consider the set $\{\mathcal{J}^\pm\}$ of all limit functions of the sequence $\{I_k^\pm\}$. The set $\{\mathcal{J}^\pm\}$ consists of contractive holomorphic functions in $\mathbb{D}_\pm$. In view of (5.15), the set $\{\mathcal{J}^\pm\}$ contains not only the identically zero function. Consider the inner-outer factorization $C = I_C^\pm E_C^\pm$ for each nonzero function C from $\{\mathcal{J}^\pm\}$.

Definition 5.3. Let

$$\mathcal{J} := \{I_k^+, I_k^+\}_{k \in \mathbb{N}} \qquad (5.16)$$

be a sequence of pairs of inner functions in $\mathbb{D}_+$ and $\mathbb{D}_-$ such that the condition (5.15) are satisfied. Then *the plus/minus denominator $I_{\mathcal{J}}^\pm$ corresponding to the sequence $\mathcal{J}$ is defined to be the greatest common divisor of the set $\{I_C^\pm\}$ of inner factors of all the nonzero limit functions C of the sequence $\{I_k^\pm\}$.* (The plus/minus denominator $I_{\mathcal{J}}^\pm$ is an inner function in $\mathbb{D}_\pm$.)

Now we need the notion of *the lower limit of a sequence of subspaces*. Let $\mathfrak{X}$ be a metric vector space, i.e. a vector space equipped with a translation invariant metric ρ such that the linear operations in $\mathfrak{X}$ are continious with respect to the metric ρ.

Definition 5.4. (See [Nik2, Lecture II, item 2.) Let $\mathfrak{X}$ be a metric vector space and let $\{E_k\}_{k\in\mathbb{N}}$ be a sequence of its subspaces. By definition, *the lower limit of this sequence* is the set $\underline{\lim}_k E_k$,

$$\underline{\lim}_k E_k := \{x \in \mathfrak{X} : \lim_{k\to\infty} dist(x, E_k) = 0\} =$$
$$= \{x \in \mathfrak{X} : \exists \{x_k\}_{k\in\mathbb{N}} : x_k \in E_k, \lim_{k\to\infty} \rho(x, x_k) = 0\}$$

It is clear that $\underline{\lim}_k E_k$ is a (closed) subspace of $\mathfrak{X}$ and that if all the E_k are invariant for a given operator $\mathbf{S}$, then the lower limit $\underline{\lim}_k E_k$ will also be invariant for $\mathbf{S}$.

Now we are able to formulate the generalization of G.Ts. Tumarkin theorem using the language of spectral function theory.

Theorem 5.2. *Let* $\mathfrak{I} = \{I_k^+, I_k^-\}_{k\in\mathbb{N}}$ *be a given sequence of pairs of inner functions in* $\mathbb{D}_+$ *and* $\mathbb{D}_-$, *respectively, which satisfies condition (5.15). Let* L_w^p *be a given weighted space, where* $p \in (0, \infty)$, $w \in \mathcal{W}$, *and let*

$$E_k := \mathbf{K}_w^p((I_k^-)^\#, (I_k^+)^\#) \qquad (k \in \mathbb{N}) \tag{5.17}$$

be a sequence of subspaces of the space L_w^p *(see Definition 5.2). Then the lower limit* $\underline{\lim}_k E_k$ *of the sequence* $\{E_k\}$ *(with respect to the* L_w^p-*metric) is contained in the space* $PCH_w^p(I_{\mathfrak{I}}^+, I_{\mathfrak{I}}^-)$ *(see Definition 4.2), where* $I_{\mathfrak{I}}^\pm$ *are plus/minus denominators corresponding to the sequence* $\mathfrak{I}$:

$$\underline{\lim}_k E_k \subseteq PCH_w^p(I_{\mathfrak{I}}^+, I_{\mathfrak{I}}^-). \tag{5.18}$$

The formulation of Theorem 5.2 was motivated by Lemma 5.3. We will not give a proof of Theorem 5.2 here; it can be obtained by methods which are similar to those used by G.Ts. Tumarkin ([Tum7], [Tum9]). Theorem 5.2 can be generalized to vector-valued functions ($p = 2$, w is a given nonnegative summable matrix-valued function satisfying the Szegö condition $\int_{\mathbb{T}} (\ell n[det\, w(t)]) m(dt) > -\infty$; $I_k^\pm$ are matrix-valued inner function in $\mathbb{D}_\pm$). In the future, we are planing to devote a special series of articles to such "vectorial" generalizations.

Remark 5.2. In the formulation of Theorem 5.2 we have used the notion of "the greatest common divisor of a given set of inner functions". There exist various ways to obtain the greatest common divisor. Given a family $\{I_\alpha^+\}_\alpha$ of inner function in $\mathbb{D}_+$ (α is a parameter which "enumerates" the family), we consider the Riesz-Nevanlinna-Smirnov factorization

$$I_\alpha^+(z) = \Big(\prod_j \frac{a_j^{(\alpha)} - z}{1 - \overline{a_j^{(\alpha)}}z} \cdot \frac{|a_j^{(\alpha)}|}{a_j^{(\alpha)}}\Big) \exp\{-\int_\mathbb{T} \frac{t+z}{t-z}\sigma_s^{(\alpha)}(dt)\} \qquad (z \in \mathbb{D}_+)$$

for each inner function I_α^+, where the $\{a_j^{(\alpha)}\}_j$ are the zeros of I_α^+, and $\sigma_s^{(\alpha)}(dt)$ is a nonnegative singular measure. Then we define

$$I^+(z) := \Big(\prod_j \frac{a_j - z}{1 - \overline{a_j}z} \cdot \frac{|a_j|}{a_j}\Big) \exp\{-\int_\mathbb{T} \frac{t+z}{t-z}\sigma_s(dt)\} \qquad (z \in \mathbb{D}_+)$$

where the a_j are the "common points" in the family $\{a_j^{(\alpha)}\}_\alpha$ and the measure $\sigma_s(dt)$ is the lower envelope of the family of measures $\{\sigma_s^{(\alpha)}(dt)\}$. The function I^+ turns out to be the greatest common divisor of the family $\{I_\alpha^+\}_\alpha$. (See a similar construction of the greatest common divisor in [Gar, Ch.II, Corollary 7.2].)

It is also possible to introduce a plus denominator for a given sequence $\{I_k^+\}_k$ of inner functions in $\mathbb{D}_+$ in a more "geometric" way by using the theory of shift invariant subspaces. (See an outline of such an approach in [Nik2, Lecture 1, item 3].) As a rule, the theory of shift invariant subspaces in *the Hardy class* $H^2(\mathbb{D}_+)$ is used for this purposes. However, it is more natural to use the theory of shift invariant subspaces in the weighted Hardy class $H_w^p(\mathbb{D}_+)$. (For a weight function $w \in W$, the weighted Hardy class $H_w^p(\mathbb{D}_+)$ is defined as the set of all functions f which (1) are holomorphic in $\mathbb{D}_+$, (2) belong to the V.I. Smirnov class $N_*(\mathbb{D}_+)$, and (3) satisfy the condition $\int_\mathbb{T} |f(t)|^p w(t)m(dt) < \infty$. The ordinary Hardy class corresponds to $w(t) \equiv 1$.) Every nonzero shift-invariant subspace of the space $H_w^p(\mathbb{D}_+)$ has the form $\Theta H_w^p(\mathbb{D}_+)$, where Θ is an inner function in $\mathbb{D}_+$. The operator $f(z) \rightarrow f(z)\Phi^+(z)$ (Φ^+ was defined in (4.13)) maps the space $H_w^p(\mathbb{D}_+)$ isometrically onto the space $H^p(\mathbb{D}_+)$ and interwines the shift operators in these spaces. Let $\{I_k^+\}_{k \in \mathbb{N}}$ be a sequence of inner functions in $\mathbb{D}_+$ and let $E_k := I_k^+ H_w^p(\mathbb{D}_+)$ be the associated shift-invariant subspaces of the space $H_w^p(\mathbb{D}_+)$. Then $E := \lim_k E_k$ (see Definition 5.4) is a shift-invariant subspace of the space $H_w^p(\mathbb{D}_+)$. Therefore it has the form $E = I^+ H_w^p(\mathbb{D}_+)$, where I^+ is an inner function in $\mathbb{D}_+$. The function I^+ proves to be a plus-denominator of the sequence $\{I_k^+\}_k$.

6. Formulation of the Main Approximation Theorem.

The next theorem, Theorem 6.1, is a converse to Theorem 5.1. The special case with $w(t) \equiv 1$ and $1 \leq p < \infty$ was established earlier by G.Ts. Tumarkin. The general version stated below, is proved here for the first time.

Theorem 6.1. *Let a space $PCH_w^p(I^+, I^-)$ be given, where $p \in (0, \infty)$, $w \in \mathcal{W}$ and $I^{\pm}$ are inner functions in $\mathbb{D}_{\pm}$. (The space $PCH_w^p(I^+, I^-)$ is defined in Definition 4.2.)*

Then there exists a sequence $\mathcal{J} = \{B_k^+, B_k^-\}_k \in \mathbb{N}$ of pairs of finite Blaschke products ($B_k^{\pm}$ is a Blaschke product in $\mathbb{D}_{\pm}$) with the following properties:

i). $\lim_{k\to\infty} B_k^{\pm}(z) = I^{\pm}(z), \qquad (z \in \mathbb{D}_{\pm})$;

    (In particular, condition (5.15) is satisfied and the given pair (I^+, I^-) of inner function is a pair of plus- and minus denominators corresponding to the sequence $\mathcal{J}$.)

ii). *The space PCH_w^p is the lower limit of the sequence of subspaces*

$$\mathbf{K}_w^p((B_k^-)^{\#}, (B_k^+)^{\#}) \text{ of the space } L_w^p.$$

In view of the description of the space $\mathbf{K}_w^p((B_k^-)^{\#}, (B_k^+)^{\#})$ given in Lemma 5.3, Theorem 6.1 is a theorem on the weighted approximation of pseudocontinuable functions from the space $PCH_w^p(I^+, I^-)$ by rational functions with a prescribed table of poles. We emphasize that <u>one and the same</u> table $\mathcal{J}$ enables us to approximate <u>all</u> functions f from the space $PCH_w^p(I^+, I^-)$.

The special case of this theorem in which $I^{\pm} \equiv 1$ was established in item 6 of [Kats2]; see the "Fundamental Theorem" there. Another special case of Theorem 6.1 ($w(t) \equiv 1$, $I^{\pm}$ arbitrary inner functions) was proved by G.Ts. Tumarkin. (See [Tum7] for $p \in [1, \infty)$ and [Tum9] for $p \in (0, \infty)$.) We emphasize that G.Ts. Tumarkin considered an "individual approximation" only: he considered an arbitrary <u>fixed</u> function $f \in PCH_1^p(I^+, I^-)$ and proved that <u>for this f</u> there exist a table $\mathcal{J} = \mathcal{J}_f$ (depending on f) of the form (1.1) - (1.2) (or, what is the same, a sequence $\mathcal{J} = \{B_k^+, B_k^+\}_{k \in \mathbb{N}}$ of pairs of finite Blaschke products in $\mathbb{D}_{\pm}$) which satisfies the following two conditions:

i). The functions I^+, I^- are the plus-denominator and the minus- denominator of the table $\mathcal{J}$;

ii). The function f belongs to the lower limit $\underline{\lim}_k E_k$ of the sequence $\{E_k\}_k$ of spaces $E_k := \mathbf{K}_w^p((B_k^-)^{\#}, (B_k^+)^{\#})$.

7. A Fundamental Approximation Lemma.

The following approximation lemma is the core of what follows. This lemma has an *'individual'* character. The objective is to construct a table of poles for a sequence of rational functions which tends in the L_w^p-metric to a *given* function in the space $PCH_w^p(I^+, I^-)$. We assume that both I^+ and I^- are Blaschke products. This particular case serves as a basis for more general considerations; it contains the main analytic difficulties, which will be resolved by invoking methods drawn from the theory of analytic matrix-valued functions.

Notation: Given a function h which is meromorphic in the open (disconnected) set $D_+ \cup D_-$, we denote the set of all the zeros [resp. poles] of h located in $D_+ \cup D_-$ by $\mathcal{Z}_h$ [resp. $\mathcal{P}_h$]. We also set $\mathcal{Z}_h^{\pm} := \mathcal{Z}_h \cap D_{\pm}$ and $\mathcal{P}_h^{\pm} := \mathcal{P}_h \cap D_{\pm}$.

The Fundamental Approximation Lemma. *Let a space $PCH_w^p(B^+, B^-)$ be given, where $p \in (0, \infty)$, w is a summable weight function, and $B^{\pm}$ are given Blaschke products in $D_{\pm}$. (These Blaschke products may be infinite or finite. The case $B^+ \equiv 1$ or $B^- \equiv 1$ is permitted). Let f be a given function from the space $PCH_w^p(B^+, B^-)$.*

Then, for each given fixed number ε, $\varepsilon > 0$, there exists a rational function R and a set A, $A \subseteq D_+ \cup D_-$, ($A_+ := A \cap D_+$, $A_- := A \cap D_-$) which possess the following properties:

i).   *The inequality*

$$\int_{\mathbb{T}} |f(t) - R(t)|^p w(t) m(dt) < \varepsilon^p \tag{7.1}$$

   holds.

ii).   *The set A is discrete in $D_+ \cup D_-$, and the condition*

$$\sum_{z_k \in A_+} (1 - |z_k|) + \sum_{z_k \in A_-} (1 - \frac{1}{|z_k|}) < \varepsilon \tag{7.2}$$

   is fulfilled.

iii).   *The set $\mathcal{P}_R$ of poles of R (counting multiplicities) satisfies the inclusion*

$$\mathcal{P}_R \subseteq \mathcal{E} := \mathcal{Z}_{B^+} \cup \mathcal{Z}_{B^-} \cup A \tag{7.3}$$

More formally,

$$Rstr._{D_{\pm}}(RB_{\mathcal{E}}^{\pm}) \in H^{\infty}(D_{\pm}), \tag{7.4}$$

where $B_{\mathcal{E}}^{\pm} := B^{\pm} B_A^{\pm}$ and $B_A^{\pm}$ are Blaschke products in $D_{\pm}$: $\mathcal{Z}_{B_A^{\pm}} = A \cap D_{\pm}$. (In other words, $B_{\mathcal{E}}^{\pm}$ are Blaschke products in $D_{\pm}$ with zero sets $\mathcal{E}_{\pm} := \mathcal{Z}_{B^{\pm}} \cup A_{\pm}$.)

Remark 7.1. Even though $\mathcal{P}_f \subseteq \mathcal{Z}_{B^+} \cup \mathcal{Z}_{B^-}$ for $f \in PCH_w^p(B^+, B^-)$, it is impossible to approximate the given pseudocontinuable function $f \in PCH_w^p(B^+, B^-)$ by rational functions R whose poles are contained in the set $\mathcal{Z}_{B^+} \cup \mathcal{Z}_{B^-}$ only. We have to extend the set $\mathcal{Z}_{B^+} \cup \mathcal{Z}_{B^-}$ and to consider rational functions R whose poles satisfy condition (7.3) where an "additional" set $\mathcal{A} \subseteq \mathbb{D}_+ \cup \mathbb{D}_-$ satisfies (7.2) with an arbitrary small (but nonzero) fixed $\varepsilon > 0$. Condition (7.2) means that the set $\mathcal{A}$ consist of points which are positioned very close to the unite circle $\mathbb{T}$.

Remark 7.2. A phenomenon similar to the phenomenon discussed above in Remark 7.2 occurs in a weight approximation problem for entire functions of exponential type. Let $w : \mathbb{R} \to [0, +\infty]$ be a weight function which is summable and satisfies the Szegö condition $\int_{\mathbb{R}} [\ell n \, w(t)] (1+t^2)^{-1} dt > -\infty$. Let f be an entire function whose exponential type does not exceed σ : $\overline{\lim} \frac{\ell n |f(z)|}{|z|} \leq \sigma$, and assume f is weight square summable: $\int_{\mathbb{R}} |f(t)|^2 w(t) dt < \infty$. For such a function f it is possible to define its spectrum $spec_f$ in a natural way. $Spec_f$ is a closed set, $spec_f \subseteq [-\sigma, \sigma]$. The objective is to approximate the function f by trigonometric polynomials $T(t) = \sum_{\lambda} c_{\lambda} e^{i\lambda t}$ with respect to L^p_w norm. (Each sum T contains finitely many nonzero terms only.) This is nothing but S. Bernstein's problem on trigonometric approximation. (The trigonometric Bernstein approximation problem was discussed in the papers of P. Koosis [Koo1], N. Levinson and H.P. McKean [L-McK] (see item 13 there), Yu.I. Ljubarskiĭ [Lju] and V.E. Katsnelson [Kats1].) The basic problem is to evaluate the infimum $\inf_T \int_{\mathbb{R}} |f(t) - T(t)|^p w(t) dt$ under various restrictions on the spectra of the approximating polynomials T. If T runs over the set of all trigonometric polynomials whose spectrum is contained in $[-\sigma, \sigma]$ (i.e. $T(t) = \sum_{\lambda} c_{\lambda} e^{i\lambda t}$ with $\lambda \in [-\sigma, \sigma]$), then the infimum may be positive and may be equal to zero. (It depends on f and w.) However, if T runs over the set of all trigonometric polynomials whose spectrum is contained in an "extended" set $[-(\sigma + \varepsilon), (\sigma + \varepsilon)]$ ($\varepsilon > 0$ is an arbitrary fixed number), then the infimum is equal to zero.

Proof of the Fundamental Approximation Lemma. We divide the proof into two steps. At the first step we approximate an *arbitrary* given function $f \in PCH^p_w(B^+, B^-)$ (with respect to the L^p_w-topology) by a pseudocontinuable function g which is *bounded* on $\mathbb{T}$. The price which we have to pay for such an approximation is a slight *extension* of the "spectrum" of the function f. The approximating function g may have an additional set $\mathcal{A}$ of poles in $\mathbb{D}_+ \cup \mathbb{D}_-$, i.e., in general, the approximating function g will not belong to the original space $PCH^p_w(B^+, B^-)$. It will, however, belong to the "extended" space $PCH^p_w(B^+_{\mathcal{E}}, B^-_{\mathcal{E}})$, where $B^{\pm}_{\mathcal{E}} := B^{\pm} B^{\pm}_{\mathcal{A}}$, $B^{\pm}_{\mathcal{A}}$ are Blaschke products in $\mathbb{D}_{\pm}$ whose poles are positioned very close to the unit circle $\mathbb{T}$. At the second step we approximate this g by rational functions. In general, we follow the scheme developed in our paper [Kats2] (see especially item 7 there). However, in [Kats2] we considered the approximation of pseudo-continuable functions which are holomorphic (not meromorphic) in $\mathbb{D}_+ \cup \mathbb{D}_-$. Therefore we have to modify the reasoning presented there.

Step 1. Let f be an arbitrary given function from $PCH_w^p(B^+, B^-)$ and let

$$f_\alpha(z) := \frac{f(z)}{1 + \alpha h(z)} \qquad (z \in \mathbb{D}_+ \cup \mathbb{D}_-), \tag{7.5}$$

where α is a positive number and h is pseudocontinuable function satisfying the following two conditions:

i). $\quad h_\pm \in N_*(\mathbb{D}_\pm), \qquad h_+ = (h_-)^\#,$ \hfill (7.6)

where, as usually, $h_\pm := Rstr._{\mathbb{D}_\pm} h.$

ii). $\quad h(t) \geq |f(t)|^2 \qquad$ (a. e. on $\mathbb{T}$) \hfill (7.7)

Such a function h will be constructed later. From (7.5) and (7.7) it follows that

$$|f_\alpha(t)| \leq \frac{1}{2\sqrt{\alpha}} \qquad (t \in \mathbb{T}, \ \alpha > 0), \tag{7.8}$$

i.e. each function f_α from the family (7.5) is bounded on $\mathbb{T}$ by a constant (depending on α). On the other hand, from (7.5) we infer that

$$|f_\alpha(t)| \leq |f(t)| \qquad (t \in \mathbb{T}, \ \alpha > 0), \tag{7.9}$$

i.e., the family $\{f_\alpha\}_{\alpha \in (0,\infty)}$ has a common majorant. Therefore, since $f_\alpha \to f(t)$ as $\alpha \downarrow 0$, and $\int_\mathbb{T} |f(t)|^p w(t) m(dt) < \infty$, the Lebesgue dominated convergence theorem implies that

$$\int_\mathbb{T} |f(t) - f_\alpha|^p w(t) m(dt) \to 0 \qquad (\alpha \downarrow 0). \tag{7.10}$$

For $\alpha \in (0, \infty)$, set

$$d_\alpha(z) = 1 + \alpha h(z) \qquad (z \in \mathbb{D}_+ \cup \mathbb{D}_-) \tag{7.11}$$

As the function h is "symmetric" i.e., $h = h^\#$, the function d_α is "symmetric" also: $d_\alpha = (d_\alpha)^\#$. In view of (7.6),

$$Rstr._{\mathbb{D}_\pm} d_\alpha \in N_*(\mathbb{D}_\pm). \tag{7.12}$$

The zero set $\mathcal{Z}_{d_\alpha}$ is symmetric with respect to $\mathbb{T}$. Applying Jensen's inequality to $Rstr._{\mathbb{D}_+} d_\alpha$, we obtain

$$\sum_{z_k(\alpha) \in \mathbb{D}_+} \ln \frac{1}{|z_k(\alpha)|} \leq \int_\mathbb{T} [\ln |d_\alpha(t)|] m(dt) - \ln |d_\alpha(0)|, \tag{7.13}$$

where the sum on the left-hand side of (7.13) is taken over the set $\{z_k(\alpha)\}$ of all zeros of the function d_α which are contained in the unit disc $\mathbb{D}_+$. (The version of Jensen's inequality where the integral on the right-hand side of (7.13) is taken over $\mathbf{T}$ is not generally true for all functions belonging to the holomorphic Nevanlinna class $N(\mathbb{D}_+)$, but it is true for all functions which are members of the holomorphic Smirnov class $N_*(\mathbb{D}_+)$.) From (7.11) it follows that $\ell n|d_\alpha(z)| \to 0$ $(\alpha \downarrow 0)$ for every $z \in \mathbb{D}_+$ and $\ell n|d_\alpha(t)| \to 0$ $(\alpha \downarrow 0)$ for almost very $t \in \mathbf{T}$. In view of (7.6), $\int_{\mathbf{T}} \ell n(1 + h(t))m(dt) < \infty$. Therefore, since $0 \le \ell n|d_\alpha(t)| \le \ell n(1 + h(t))$ $(t \in \mathbf{T}, \ \alpha \in (0,1])$, the Lebesgue dominated convergence theorem guarantees that $\int_{\mathbf{T}} \ell n|d_\alpha(t)|m(dt) \to 0$ $(\alpha \downarrow 0)$. Now from (7.13) we infer that

$$\lim_{\alpha \downarrow 0} \left(\sum_{z_k(\alpha)\in\mathbb{D}_+} \ell n \frac{1}{|z_k(\alpha)|} \right) = 0 \tag{7.14}$$

Let ε be an arbitrary given positive number. By virtue of (7.10) and (7.14), there exists a number $\lambda = \lambda(\varepsilon, f) > 0$ such that for every $\alpha \in (0, \lambda)$, the two inequalities

$$\int_{\mathbf{T}} |f(t) - f_\alpha(t)|^p w(t)m(dt) < \varepsilon \tag{7.15}$$

and

$$\sum_{z_k(\alpha)\in\mathbb{D}_+} [1 - |z_k(\alpha)|] + \sum_{z_k(\alpha)\in\mathbb{D}_-} [1 - |z_k(\alpha)|^{-1}] < \varepsilon \tag{7.16}$$

hold. The sums in (7.16) are taken over all the zeros of the function d_α (the latter being defined in (7.11))) which are located in $\mathbb{D}_+ \cup \mathbb{D}_-$. Let $d_\alpha^+ = I_\alpha^+ E_\alpha^+$ be an inner-outer factorization of the function $d_\alpha^+ := Rstr._{\mathbb{D}_+} d_\alpha$, and let $I_\alpha^+ = B_\alpha^+ S_\alpha^+$ be a multiplicative decomposition of the inner function I_α^+ into the product of a Blaschke component B_α^+ and a singular component S_α^+. Now we show that one can choose a number $\alpha \in (0, \lambda)$ such that $S_\alpha^+ \equiv 1$ i.e., the inner component I_α^+ is a pure Blaschke product. In other words, one can choose a number $\alpha \in (0, \lambda)$ such that the function d_α^+ is representable as

$$d_\alpha^+ = B_\alpha^+ E_\alpha^+, \tag{7.17$^+$}$$

where B_α^+ is a Blaschke product and E_α^+ is an outer function in $\mathbb{D}_+$ (i.e. $E_\alpha^+ \in N_*(\mathbb{D}_+)$, $(E_\alpha^+)^{-1} \in N_*(\mathbb{D}_+)$). We rely on a theorem of O. Frostman or, more precisely, on its generalization given by W. Rudin. (See the article [Rud1], or Section 3.6 of the monograph [Rud2].) Such a generalization was also obtained independently by S.A. Vinogradov, [Vin]. The fact that a function d_α^+ in the Smirnov class $N_*(\mathbb{D}_+)$ has an inner-outer factorization of the form (7.17^+) (where the inner component of the factorization is a Blaschke product) is equivalent to the limit relation

$$\lim_{\rho\uparrow1} \int_{\mathbb{T}} \ell n|d_\alpha(\rho t)|m(dt) = \int_{\mathbb{T}} \ell n|d_\alpha(t)|m(dt)$$

or (see (7.11))

$$\lim_{\rho\uparrow1} \int_{\mathbb{T}} [\ell n|\frac{1}{\alpha} + h(\rho t)|]m(dt) = \int_{\mathbb{T}} [\ell n|\frac{1}{\alpha} + h(t)|]m(dt) \tag{7.18}$$

In Theorem 4 of Rudin's paper [Rud1] the following result is proved. *Let h be a function which belongs to the holomorphic Smirnov class $N_*(\mathbb{D}_+)$ and let K be an arbitrary compact subset of $\mathbb{C}$ with positive logarithmic capacity. Then there exists $\lambda \in K$ such that for the function $\lambda + h$ the condition*

$$\lim_{\rho\uparrow1} \int_{\mathbb{T}} [\ell n|\lambda + h(\rho t)|]m(dt) = \int_{\mathbb{T}} [\ell n|\lambda + h(t)|]m(dt) \tag{7.19}$$

holds. (This theorem is reproduced in his book [Rud2, Section 3.6].) Since every interval of the real axis has positive logarithmic capacity, there exists a number λ in each interval such that (7.19) holds. Letting $\lambda := \frac{1}{\alpha}$, we conclude that the set of all real α which satisfy the limit relation (7.17) is dense in $\mathbb{R}$. In particular, such a number α can be found in every interval $(0, \delta)$. Since the function d_α is "symmetric", i.e. $(d_\alpha)^\# = d_\alpha$, the representation (7.17^+) is equivalent to the representation

$$d_\alpha^- = B_\alpha^- E_\alpha^- \tag{7.17^-}$$

$(d_\alpha^- := Rstr._{\mathbb{D}_-} d_\alpha)$ where B_α^- is a Blaschke product in $\mathbb{D}_-$, and E_α^- is an outer function in $\mathbb{D}_-$, i.e. $E_\alpha^- \in N_*(\mathbb{D}_-)$, $(E_\alpha^-)^{-1} \in N_*(\mathbb{D}_-)$. (Namely, $B_\alpha^- = (B_\alpha^+)^\#$, $E_\alpha^- = (E_\alpha^+)^\#$.) W. Rudin uses the following definition: *A compact subset K of the complex plane is said to have positive logarithmic capacity if there exists a positive Borel measure $\mu \neq 0$ which is concentrated on K such that logarithmic potential $U_\mu(z) := \int_K [\log|z - \xi|]\mu(d\xi)$ is continuous at each point $z \in \mathbb{C}$.* Every interval of the real axis has positive logarithmic capacity because one can choose μ equal to the restriction of one-dimensional Lebesgue measure to the Borel subsets of this interval. One can easily check by straightforward computations that the logarithmic potential of this measure is continuous in the whole complex plane. A good exposition of logarithmic potential theory is presented in the book [VP]. In particular, see Lemma 68 on the page 84 there.) Frostman's original result which was referred to above is not presented for arbitrary functions of the Smirnov class, but only for inner functions. This was proved in O. Frostman's remarkable paper [Fr], which also contains many important application of potential theory in complex function theory. Now we choose a number $\alpha > 0$ such that both the inequalities (7.15) and (7.16) hold and the function $d_\alpha = 1 + \alpha h$ has factorization of the form (7.17$^+$) and (7.17$^-$) in $\mathbb{D}_+$ and $\mathbb{D}_-$, respectively, and set

$$g(z) := f_\alpha(z)\left(= \frac{f(z)}{d_\alpha(z)} = \frac{f(z)}{1 + \alpha h(z)}\right) \qquad (z \in \mathbb{D}_+ \cup \mathbb{D}_-) \qquad (7.20)$$

and

$$B_{\mathcal{A}}^+ := B_\alpha^+, \qquad B_{\mathcal{A}}^- := B_\alpha^-. \qquad (7.21)$$

(B_α^+ and B_α^- are the same as in (7.17$^+$) and (7.17$^-$).) Clearly $B_{\mathcal{A}}^+ = (B_{\mathcal{A}}^-)^\#$. Let

$$\mathcal{A} = \mathcal{Z}_{d_\alpha} \qquad \text{and} \qquad \mathcal{A}_\pm = \mathcal{A} \cap \mathbb{D}_\pm \qquad (7.22)$$

Then $\mathcal{A}_+ = (\mathcal{A}_-)^\#$, and condition (7.16) is the same as condition (7.2) for this set $\mathcal{A}$. Set

$$B_{\mathcal{E}}^\pm := B^\pm B_{\mathcal{A}}^\pm \qquad (7.23)$$

From (7.20), (7.17$^+$) and (7.23) it follows that

$$g_+ B_{\mathcal{E}}^+ = f^+ B^+(E_\alpha^+)^{-1} \in N_*(\mathbb{D}_+).$$

More precisely, $f_+ B^+ \in N_*(\mathbb{D}_+)$, because $f \in PCH_w^p(B^+, B^-) \subseteq PCNM(B^+, B^-)$ and $(E_\alpha^+)^{-1} \in N_*(\mathbb{D}_+)$ because E_α^+ is outer. Hence, $g_+ B_{\mathcal{E}}^+ \in N_*(\mathbb{D}_+)$. Similarly, $g_- B_{\mathcal{E}}^- \in N_*(\mathbb{D}_-)$. Since the functions f and h are pseudocontinuable, the function $g = f(1 + \alpha h)^{-1}$ is pseudocontinuable as well. Thus,

$$g \in PCNM(B_{\mathcal{E}}^+, B_{\mathcal{E}}^-). \qquad (7.24)$$

The inequality $\quad |g(t)| \leq |f(t)|\ (t \in \mathbb{T}) \quad$ guarantees that $\int_{\mathbb{T}} |g(t)|^p w(t) m(dt) \leq \int_{\mathbb{T}} |f(t)|^p w(t) m(dt) < \infty$. Hence,

$$g \in PCH_w^p(B_{\mathcal{E}}^+, B_{\mathcal{E}}^-). \qquad (7.25)$$

Moreover, inequalities (7.15) and (7.8) guarantee that

$$\int_{\mathbb{T}} |f(t) - g(t)|^p w(t) m(dt) < \varepsilon^p \qquad (7.26)$$

$$\sup_{t \in \mathbb{T}} |g(t)| < \infty \qquad (7.27)$$

respectively. Besides, it follows from (7.20), that $\mathcal{P}_g \subseteq \mathcal{P}_f \cup \mathcal{Z}_{d_\alpha}$. Therefore, since $\mathcal{P}_f \subseteq \mathcal{Z}_{B^+} \cup \mathcal{Z}_{B^-}$ and $\mathcal{Z}_{d_\alpha} := \mathcal{A}$,

$$\mathcal{P}_g \subseteq \mathcal{Z}_{B^+} \cup \mathcal{Z}_{B^-} \cup \mathcal{A}. \qquad (7.28)$$

Thus, we have constructed a pseudocontinuable function g which is meromorphic in $\mathbb{D}_+ \cup \mathbb{D}_-$, bounded on $\mathbb{T}$, and approximates the original function f with respect to the L_w^p-topology. Moreover, the function g is not only pseudocontinuable but also (7.24) holds with $B_{\mathcal{E}}^{\pm}$ defined in (7.23) where $B_{\mathcal{A}}^+$ and $B_{\mathcal{A}}^-$ are Blashke products: $\mathcal{Z}_{B_{\mathcal{A}}^+} = \mathcal{A}_+$, $\mathcal{Z}_{B_{\mathcal{A}}^-} = \mathcal{A}_-$. For the set $\mathcal{A} = \mathcal{A}_+ \cup \mathcal{A}_-$ the condition (7.2) holds.

However, the construction of the function g is *incomplete:* we have assumed the existence of a pseudocontinuable function h which satisfies the conditions (7.6) and (7.7). To construct such a function h, we start with the function

$$h_{st}(z) := f(z)(f(z))^{\#} \qquad (z \in \mathbb{D}_+ \cup \mathbb{D}_-). \qquad (7.29)$$

Clearly, h_{st} is pseudocontinuable, since f is, and

$$h_{st}(t) = |f(t)|^2 \qquad (t \in \mathbb{T}) \qquad (7.30)$$

Condition (7.30) for the function h_{st} has the desired form (7.7). If the function f is holomorphic in $\mathbb{D}_+ \cup \mathbb{D}_-$ (recall that $f_{\pm} B^{\pm} \in N_*(\mathbb{D}_{\pm})$, hence the functions $f_{\pm}$ belong to the meromorphic Smirnov class $NM_*(\mathbb{D}_{\pm})$ respectively), then the function h_{st} fulfills the condition (7.6). In this case we may set $h = h_{st}$. (In particular, we may do this if $B_{\pm} \equiv 1$. This is the construction which we used in [Kats2]. See formula (7.13) there.) However, in general h_{st} is only meromorphic (and not holomorphic) in $\mathbb{D}_+ \cup \mathbb{D}_-$. Therefore we must "remove" its poles from $\mathbb{D}_+ \cup \mathbb{D}_-$ in a way which ensures the validity of conditions (7.6) and (7.7). We shall achieve this by "sweeping" the poles of the function h_{st} from the open set $\mathbb{D}_+ \cup \mathbb{D}_-$ to boundary - the unit circle $\mathbb{T}$. It is this sweeping process that is new, in comparison with the construction in [Kats2, item 7]. Set

$$B := B_+ (B_-)^{\#} \qquad (7.31)$$

The function B is a Blaschke product in $\mathbb{D}_+$, and $\mathcal{Z}_B = \mathcal{Z}_{B^+} \cup (\mathcal{Z}_{B^-})^{\#}$. Moreover, since $f_+ B^+ \in N_*(\mathbb{D}_+)$ and $f_- B^- \in N_*(\mathbb{D}_-)$, $(h_{st})_+ B \in N_*(\mathbb{D}_+)$ and $(h_{st})_- B^{\#} \in N_*(\mathbb{D}_-)$. Thus

$$h_{st} \in PCNM(B, B^{\#}) \qquad (7.32)$$

We now describe the "sweeping process". To the end, let z_1, z_2, ... be the zeros of the Blaschke product B defined by (7.31), and we suppose for simplicity that $z_k \neq 0$. Then

$$B(z) = \prod_k \frac{z_k - z}{1 - \bar{z}_k z} \cdot \frac{|z_k|}{z_k}, \quad B^{\#}(z) = \prod_k \frac{1 - z\bar{z}_k}{z_k - z} \cdot \frac{|z_k|}{\bar{z}_k} \quad (z \in \mathbb{D}_+ \cup \mathbb{D}_-). \quad (7.33)$$

Set

$$z_k = |z_k| t_k \quad (t_k \in \mathbf{T},\ 0 < |z_k| < 1), \quad (7.34)$$

and

$$P(z) := \prod_k p_k(z) \quad (z \in \mathbb{D}_+ \cup \mathbb{D}_-). \quad (7.35)$$

where

$$p_k(z) := \frac{1}{|z_k|} \cdot \frac{z - z_k}{z - t_k}. \quad (7.36)$$

The product on the right-hand side of (7.35) converges locally uniformly in $\mathbb{D}_+ \cup \mathbb{D}_-$, since $\left| \frac{z-z_k}{z-t_k} - 1 \right| \leq (1-|z_k|)\frac{1}{|1-|z||}$ and $\sum_k (1 - |z_k|) < \infty$. Thus, the function P is holomorphic in $\mathbb{D}_+ \cup \mathbb{D}_-$.

Definition 7.1.   The function P defined by (7.35) and (7.36) is said to be *a sweeping function*.

Set

$$h(z) := h_{st}(z)P(z)P^{\#}(z) \quad (z \in \mathbb{D}_+ \cup \mathbb{D}_-). \quad (7.37)$$

Then, since $\mathcal{Z}_P = \mathcal{Z}_{B^+} \cup (\mathcal{Z}_{B_-})^{\#}$, and $\mathcal{P}_{h_{st}} \subseteq \mathcal{P}_f \cup (\mathcal{P}_f)^{\#}$, the functions $f_+ B^+$, $(f_-)^{\#}(B_-)^{\#}$ are holomorphic in $\mathbb{D}_+$, and the function h is holomorphic in $\mathbb{D}_+ \cup \mathbb{D}_-$.

The properties of the sweeping function P are characterized by the following lemma.

Lemma 7.1. *The function P defined by (7.35) and (7.36) has the following properties:*

 i). $P(z) = B(z)P^{\#}(z) \quad (z \in \mathbb{D}_+ \cup \mathbb{D}_-)$ $\qquad\qquad\qquad$ (7.38)

 ii). $|P(z)| \geq 1 \quad (z \in \mathbb{D}_-)$ $\qquad\qquad\qquad\qquad\qquad$ (7.39)

 iii). $P_{\pm} \in N_*(\mathbb{D}_{\pm})$ $\qquad\qquad\qquad\qquad\qquad\qquad\qquad$ (7.40)
  where, as usual, $P_+ = Rstr._{\mathbb{D}_+} P$, $P_- = Rstr._{\mathbb{D}_-} P$.

 iv). *The function P is pseudocontinuable, i.e.*

$$P_+(t) = P_-(t) \quad \text{a. e. on } \mathbf{T} \quad (7.41)$$

We shall prove Lemma 7.1 later. Now we obtain some consequences of it. From (7.37) and (7.38) it follows that

$$h(z) = h_{st}(z)B(z)(P^{\#}(z))^2 \qquad (z \in \mathbf{D}_+ \cup \mathbf{D}_-).$$

Therefore since $(P^{\#})_+ \in N_*(\mathbf{D}_+)$ (or equivalently $P_- \in N_*(\mathbf{D}_-)$) by (7.40) and $(h_{st}B)_+ \in N_*(\mathbf{D}_+)$ (since $(fB^+)_+ \in N_*(\mathbf{D}_+)$ and $(f^{\#}(B^-)^{\#})_+ \in N_*(\mathbf{D}_+)$ it follows that $h_+ \in N_*(\mathbf{D}_+)$, and by symmetry $(h = h^{\#})$ we obtain $h_- \in N_*(\mathbf{D}_-)$. Thus (7.6) holds for the function h defined by (7.37). The function h is pseudocontinuable as well, since h_{st} and P are. Furthermore, from (7.39) it follows that $|P(t)| \geq 1$, and $|P^{\#}(t)| \geq 1$ $(t \in \mathbf{T})$, and hence, that $h(t) \geq h_{st}(t)$ $(t \in \mathbf{T})$. Thus from (7.30) we obtain (7.7).

To finish step one of the proof of the Fundamental Approximation Lemma, it remains only to prove Lemma 7.1.

Proof of Lemma 7.1. Set $b_k(z) := t_k^{-1}\frac{z_k - z}{1 - \bar{z}_k z}$. Clearly, $p_k = b_k(p_k)^{\#}$ (see (7.36)). Multiplying by k, we obtain (7.38). It is not difficult to check directly that the function p_k is expansive in $\mathbf{D}_-$: $|p_k(z)| > 1$ $(z \in \mathbf{D}_-)$. Multiplying by k, we obtain (7.39). Each of the functions $(p_k)_- := Rstr._{\mathbf{D}_-} p_k$ belongs to the Smirnov class $N_*(\mathbf{D}_-)$. It is well known that the convergent product of factors each of which is a holomorphic expansive function belonging to the Smirnov class, is a function from the Smirnov class as well. Thus, $P_- \in N_*(\mathbf{D}_-)$ (or, equivalently, $(P^{\#})_+ \in N_*(\mathbf{D}_+)$). As $|B(z)| \leq 1$ $(z \in \mathbf{D}_+)$, we obtain from (7.38) that $P_+ \in N_*(\mathbf{D}_+)$. Thus, (7.40) is proved.

Set

$$s_k(z) := (p_k(z))^{-1} = |z_k|\frac{z - t_k}{z - z_k} \qquad (z \in \mathbf{C})$$

and

$$S_n(z) = \prod_{1 \leq k \leq n} s_k(z) \qquad (z \in \mathbf{C})$$

. As $|s_k(z)| < 1$ $(z \in \mathbf{D}_-)$ (it is the same that $|p_k(z)| > 1$ $(z \in \mathbf{D}_-)$), the inequalities

$$1 > |S_1(z)| > |S_2(z)| > \cdots > |S_n(z)| > \cdots > |S(z)| \qquad (z \in \mathbf{D}_-) \qquad (7.42)$$

hold. Moreover,

$$\lim_{n \to \infty} S_n(z) = S(z) \qquad (z \in \mathbf{D}_-) \qquad (7.43)$$

From (7.42) and (7.43) it follows that

$$\int_{\mathbf{T}} |S_n(t) - S(t)|^2 m(dt) \leq \int_{\mathbf{T}} |1 - S_-(t)(S_n(t))^{-1}|^2 m(dt) \leq 2 - 2Re\frac{S(\infty)}{S_n(\infty)},$$

and hence

$$\lim_{n\to\infty} \int_{\mathbb{T}} |S_n(t) - S_-(t)|^2 m(dt) = 0 \qquad (7.44)$$

Clearly $(S_n)_+(t) = (S_n)_-(t) = S_n(t)$ $(t \in \mathbb{T})$, since S_n is rational. Multiplying the identities $b_k s_k = (s_k)^\#$ by k, we obtain the identity

$$B_n(z)S_n(z) = (S_n)^\#(z) \qquad\qquad (z \in \mathbb{C}), \qquad (7.45)$$

where

$$B_n(z) := \prod_{1 \le k \le n} b_k(z).$$

The limit identity

$$B(z)S(z) = S^\#(z) \qquad\qquad (z \in \mathbb{D}_+ \cup \mathbb{D}_-) \qquad (7.46)$$

holds as well (however, only for $z \in \mathbb{D}_+ \cup \mathbb{D}_-$). Since $\int_{\mathbb{T}} |B_n(t) - B(t)|^2 m(dt) \to 0$ $(n \to \infty)$, $|B(t)| = 1$, and $|S_n(t)| \le 1$ $(t \in \mathbb{T})$, it follows from (7.44), that $\lim_{n\to\infty} \int_{\mathbb{T}} |B_n(t)S_n(t) - B(t)S_-(t)|^2 m(dt) = 0$, or, upon taking in account (7.45) and (7.46),

$$\lim_{n\to\infty} \int_{\mathbb{T}} |(S_n)^\#(t) - (S^\#)_-(t)|^2 m(dt) = 0.$$

Together with the identity $|(S_n)^\#(t) - (S^\#)_-(t)| = |S_n(t) - S_+(t)|$ $(t \in \mathbb{T})$, this implies further that

$$\lim_{n\to\infty} \int_{\mathbb{T}} |S_n(t) - S_+(t)|^2 m(dt) = 0 \qquad (7.47)$$

From (7.44) and (7.47) it follows that $S_+(t) = S_-(t)$ $(t \in \mathbb{T})$. Thus, the function S is pseudocontinuable. The <u>first step</u> of the proof of the Fundamental Approximation Lemma is <u>finished</u>.

Step 2. In step 1 we have constructed a function $g \in PCNM(B_{\mathcal{E}}^+, B_{\mathcal{E}}^-)$ which is bounded on $\mathbb{T}$ and approximates the original function f with respect to the L_w^p-topology. The space $PCH_w^p(B_{\mathcal{E}}^+, B_{\mathcal{E}}^-)$ is a slight extension of the original space $PCH_w^p(B^+, B_-)$; see (7.23) and (7.2). The goal of step 2 is to approximate these functions whose poles are located in the set $\mathcal{Z}_{B_{\mathcal{E}}^+} \cup \mathcal{Z}_{B_{\mathcal{E}}^-} = \mathcal{Z}_{B^+} \cup \mathcal{Z}_{B^-} \cup \mathcal{A}$.

Let us introduce the functions

$$s(z) := B_{\mathcal{E}}^+(z)g(z) \qquad\qquad (z \in \mathbb{D}_+ \cup \mathbb{D}_-), \qquad (7.48)$$

and

$$B(z) := B_{\mathcal{E}}^+(z)(B_{\mathcal{E}}^-)^\#(z) \qquad (z \in \mathbb{D}_+ \cup \mathbb{D}_-). \qquad (7.49)$$

Both s and B are pseudocontinuable functions. Besides, the function B defined by (7.49) is a Blaschke product in $\mathbf{D}_+$, $\varkappa_B = \varkappa_{B_{\mathcal{E}}^+} \cup (\varkappa_{B_{\mathcal{E}}^-})^\#$. Therefore, since $\sup_t |s(t)| = \sup_t |g(t)| < \infty$ and $s_+ := Rstr._{\mathbf{D}_+} s \in N_*(\mathbf{D}_+)$ (since $g \in PCNM(B_{\mathcal{E}}^+, B_{\mathcal{E}}^-)$, $g_+B_{\mathcal{E}}^+ = s_+ \in N_*(\mathbf{D}_+)$), it follows from the Maximum Principle of V.I. Smirnov that

$$s_+ \in H^\infty(\mathbf{D}_+) \tag{7.50}$$

Next, as $(B_{\mathcal{E}}^-)^\# = (B_{\mathcal{E}}^-)^{-1}$, $(B_{\mathcal{E}}^+)^{-1} = (B_{\mathcal{E}}^+)^\#$, it follows that $sB^{-1} = gB_{\mathcal{E}}^-$ belongs to $N_*(\mathbf{D}_-)$ and hence

$$s_- B^{-1} \in H^\infty(\mathbf{D}_-) \tag{7.51}$$

Now we use the following

Approximation Theorem of D.Z. Arov ([Ar3]). *Let s be a pseudocontinuable function such that $s_+ \in H^\infty(\mathbf{D}_+)$. Suppose that there exists a Blaschke product B in $\mathbf{D}_+$ such that $(sB^{-1})_- \in H^\infty(\mathbf{D}_-)$ (or, what is the same, $s_- \in NM_*(\mathbf{D}_-)$).*

Then there exists a sequence $\{\wp_n\}_{n\in\mathbb{N}}$ of rational functions with the following properties:

i). *The limit relation*

$$\lim_{n\to\infty} \wp_n(t) = s(t) \tag{7.52}$$

holds a.e. on $\mathbf{T}$

ii). *The sequence $\{\wp_n\}_{n\in\mathbb{N}}$ is uniformly bounded on $\mathbf{T}$:*

$$\sup_n \sup_{t\in\mathbf{T}} |\wp_n(t)| = \sup_{t\in\mathbf{T}} |s(t)| \tag{7.53}$$

iii). *For every n, the poles of $\wp_n$ are located in the pole set of the function s: $\mathcal{P}_{\wp_n} \subseteq \mathcal{P}_s$ $(n \in \mathbb{N})$. More precisely,*

$$(\wp_n)_+ \in H^\infty(\mathbf{D}_+), \ (\wp_n B^{-1})_- \in H^\infty(\mathbf{D}_-) \tag{7.54}$$

(and hence, $\mathcal{P}_{\wp_n} \subseteq \mathbf{D}_-$).

This theorem, which is due to D.Z. Arov, lies in the core of our construction. We shall apply it to the function s is defined by (7.48). For the Blaschke product which occurs in the Arov's theorem we take the function B defined by (7.49). Let $\{\wp_n\}_{n\in\mathbb{N}}$ be a sequence of rational functions the existence of which is ensured by Arov's theorem. The sequence

$$\Psi_n := \wp_n(B_{\mathcal{E}}^+)^{-1} \tag{7.55}$$

$f(z) \rightarrow f(z)\Phi^+(z)$ (Φ^+ is uniformly bounded on $\mathbf{T}$:

$$\sup_n \sup_{t \in \mathbf{T}} |\Psi_n(t)| = \sup_n \sup_{t \in \mathbf{T}} |\wp_n(t)| = \sup_{t \in \mathbf{T}} |s(t)| = \sup_{t \in \mathbf{T}} |g(t)|, \tag{7.56}$$

and converges a.e. on $\mathbf{T}$ to the function g:

$$\lim_{n \to \infty} \Psi_n(t) = g(t) \qquad \text{(a. e. on } \mathbf{T}). \tag{7.57}$$

However, since the function $B_{\mathcal{E}}^+$ is not rational (in general), the function Ψ_n is not ratio-nal either. To obtain a rational function from the function Ψ_n, we have to modify the construction. Let $B_{\mathcal{E},k}^+$ be the k-th parthial Blaschke product which is generated by the "full" Blaschke product $B_{\mathcal{E}}^+$, and set

$$\Psi_{n,k} = \wp_n(B_{\mathcal{E},k}^+)^{-1} = \wp_n(B_{\mathcal{E},k}^+)^{\#} \qquad (n \in \mathbf{N}, \ k \in \mathbf{N}). \tag{7.58}$$

Clearly, every function $\Psi_{n,k}$ is rational. Moreover, the function $B_{\mathcal{E}}^+(B_{\mathcal{E},k}^+)^{-1} = B_{\mathcal{E}}^+(B_{\mathcal{E},k}^+)^{\#}$ is holomorphic in $\mathbf{D}_+$ and meromorphic in $\mathbf{D}_-$,

$$\mathbf{D}_- \supseteq \mathcal{P}_{B_{\mathcal{E}}^+(B_{\mathcal{E},1}^+)^{\#}} \supseteq \mathcal{P}_{B_{\mathcal{E}}^+(B_{\mathcal{E},2}^+)^{\#}} \supseteq \cdots, \quad \bigcap_k \mathcal{P}_{B_{\mathcal{E}}^+(B_{\mathcal{E},k}^+)^{\#}} = \emptyset. \tag{7.59}$$

From (7.54), (7.55) and (7.49) it follows that

$$Rstr._{\mathbf{D}_\pm}(\Psi_n B_{\mathcal{E}}^\pm) \in H^\infty(\mathbf{D}_\pm) \tag{7.60}$$

From (7.55) and (7.58) it follows that

$$\Psi_{n,k} = \Psi_n B_{\mathcal{E}}^+(B_{\mathcal{E},k}^+)^{\#}, \tag{7.61}$$

or $\Psi_{n,k}B_{\mathcal{E},k}^+ = \Psi_n B_{\mathcal{E}}^+$. From (7.60) we obtain now that $Rstr._{\mathbf{D}_+}(\Psi_{n,k}B_{\mathcal{E},k}^+) \in H^\infty(\mathbf{D}_+)$. Thus

$$Rstr._{\mathbf{D}_+}(\Psi_{n,k}B_{\mathcal{E}}^+) \in H^\infty(\mathbf{D}_+) \qquad (n \in \mathbf{N}, \ k \in \mathbf{N}) \tag{7.62}$$

On the other hand,

$$\Psi_{n,k}B_{\mathcal{E}}^- = (\Psi_n B_{\mathcal{E}}^-)(B_{\mathcal{E}}^+(B_{\mathcal{E},k}^+)^{\#})$$

In view of (7.60)

$$\mathcal{P}_{Rstr \cdot \mathbf{D}_-}(\Psi_{n,k} B_{\mathcal{E}}^-) \subseteq \mathcal{P}_{B_{\mathcal{E}}^+ (B_{\mathcal{E},k}^+)^\#} \qquad (n \in \mathbf{N}, \ k \in \mathbf{N}) \tag{7.63}$$

Since the function $(B_{\mathcal{E},k}^+)^\#$ is holomorphic in $\mathbf{D}_-$, it follows from (7.58), that

$$\mathcal{P}_{Rstr \cdot \mathbf{D}_-}(\Psi_{n,k} B_{\mathcal{E}}^-) \subseteq \mathcal{P}_{Rstr \cdot \mathbf{D}_-}(\wp_n) \qquad (n \in \mathbf{N}). \tag{7.64}$$

Compare now the inclusions (7.63) and (7.64). For each fixed n the set in the right-hand side of (7.64) is *finite* (the function $\wp_n$ is rational) and does not depend on k. On the other hand, the sets on the right-hand side of (7.63) decrease as k increases and has an empty intersection: (7.59). Therefore,

$$Rstr \cdot \mathbf{D}_-(\Psi_{n,k} B_{\mathcal{E}}^-) \in H^\infty(\mathbf{D}_-) \qquad (n \in \mathbf{N}; \ k > K(n), \ K(n) < \infty) \tag{7.65}$$

It is well known that there exists a subsequence $\{B_{\mathcal{E},k_n}^+\}_n$ of the sequence $\{B_{\mathcal{E},k}^+\}_k$ which converges to $B_{\mathcal{E}}^+$ a. e. on $\mathbf{T}$ (since $\int_{\mathbf{T}} |B_{\mathcal{E}}^+(t) - B_{\mathcal{E},k}^+(t)|^2 m(dt) \to 0$ as $k \to \infty$). Increasing k_n if necessary, we can assume that $k(n) > K(n)$ ($K(n)$ is defined in (7.65)). Now set

$$R_n := \wp_n(B_{\mathcal{E},k_n}^+)^\# = \Psi_{n,k_n} \qquad (n \in \mathbf{N}). \tag{7.66}$$

Every function R_n is rational. In view of (7.62) and (7.65),

$$Rstr \cdot \mathbf{D}_\pm (R_n B_{\mathcal{E}}^\pm) \in H^\infty(\mathbf{D}_\pm) \tag{7.67}$$

In other words,

$$R_n \in PCNM(B_{\mathcal{E}}^+, B_{\mathcal{E}}^-) \qquad (n \in \mathbf{N}) \tag{7.68}$$

Therefore, since $\wp_n(t) \to s(t)$ and $B_{\mathcal{E},k_n}^+(t) \to B_{\mathcal{E}}^+(t)$ as $k \to \infty$ a. e. on $\mathbf{T}$,

$$\lim_{n \to \infty} R_n(t) = g(t) \qquad (\text{a. e. on } \mathbf{T}) \tag{7.69}$$

Since $\sup_n \sup_{t \to \infty} |R_n(t)| = \sup_{t \in \mathbf{T}} |g(t)| < \infty$, we obtain that

$$\lim_{n \to \infty} \int_{\mathbf{T}} |g(t) - R_n(t)|^p w(t) m(dt) = 0 \tag{7.70}$$

(by Lebesgue's dominated convergence theorem). Thus, from (7.26) and (7.70) it follows that

$$\overline{\lim}_{n \to \infty} \int_{\mathbf{T}} |f(t) - R_n(t)|^p w(t) m(dt) < \varepsilon^p$$

Now, upon setting $R = R_n$, where n is a fixed index which is sufficiently large, we obtain inequality (7.1).

The Fundamental Approximation Lemma is proved.   □

Acknowlegements:

I would like to thank my wife Clara Katsnelson for the high quality typing of this paper. I would like to thank Harry Dym for his help in proof-reading this paper.

References

[Ar1]   AROV, D.Z.: *Darlington's method for dissipative systems* (in Russian). Doklady Akademii Nauk SSSR **201** (1971), 559 - 562. Engl. transl. in: Soviet Physics - Doklady **16** (1972), 954 - 956. **MR 55**#1127.

[Ar2]   AROV, D.Z.: *Realisation of matrix-valued functions according to Darlington* (in Russian). Izv. Akad. Nauk SSSR. Ser. Mat. **37** (1973), 1299 - 1331. Engl. transl. in: Math. USSR Izvestija **7** (1973), 1295 - 1326. **MR 50**#10287.

[Ar3]   AROV, D.Z.: *Approximating characteristic of functions of class* BΠ (in Russian). Funkcional Analiz i Ego Prilozhenija **12** (1978), 2, 70 - 71. Engl. transl. in: Functional analysis and its application **12** (1978), 133 - 134. **MR 58**#11420.

[Ar4]   AROV, D.Z.: *On regular and singular J-inner matrix-functions and related extrapolation problems* (in Russian). Funkcional'nii Analiz i Ego Prilozhenija **22** (1988), 57 - 59. Engl. transl. in: Functional analysis and its application **22** (1988), 46 - 48. **MR 89d**:47082.

[Ar5]   AROV, D.Z.: *Gamma-generating matrices, j-inner matrix functions and related extrapolation problems I* (in Russian). Teorija Funkcii, Funkcional'nii Analiz i Ikh Prilozhenija **51** (1989), 61 - 67. Engl. transl. in: Journal of Soviet Mathematics **52** (1990), 3487 - 3491. **MR 92i**:30034.

[Ar6]   AROV, D.Z.: *Regular J-inner matrix-functions and related continuation problems.* In: Operator Theory: Advances and Applications, vol. **43**. Basel: Birkhäuser 1990, 63 - 87.

[Ar7]   AROV, D.Z.: *Linear stationary passive systems with loss* (in Russian). Thesis for a Doctor's degree. Odessa 1985.

[AFK1]   AROV, D.Z., FRITZSCHE, B., and B. KIRSTEIN: *On some completion problems for various subclasses of* j_{pq}- *inner functions.* Zeitschr. für Analysis und ihre Anwend. **11** (1992), 489 - 508.

[AFK2]   AROV, D.Z., FRITZSCHE, B., and B. KIRSTEIN: *Completion problems for* j_{pq}-*inner functions. I.* Integral Equations and Operator Theory **16** (1993), 155 - 185.

[AFK3]   AROV, D.Z., FRITZSCHE, B., and B. KIRSTEIN: *Completion problems for j_{pq}-inner functions. II.* Integral Equations and Operator Theory **16** (1993), 453 - 495.

[DH]   DOUGLAS, R.G., and J.W. HELTON: *Inner dilation of analytic matrix functions and Darlington synthesis.* Acta Sci. Math. (Szeged) **34** (1973), 61 - 67. **MR 48**#900.

[DSS1]   DOUGLAS, R.G., SHAPIRO, H.S., and A.L. SHIELDS: *On cyclic vectors of the backward shift.* Bull. Amer. Math. Soc. **73** (1967), 156 - 159. **MR 34**#3316.

[DSS2]   DOUGLAS, R.G., SHAPIRO, H.S., and A.L. SHIELDS: *Cyclic vectors and invariant subspaces for the backward shift operator.* Ann. Inst. Fourier (Grenoble) **20** (1970), fasc. 1, 37 - 76. **MR 42**#5508.

[Dur]   DUREN, P.L.: *Theory of H^p Spaces.* New York and London: Academic Press 1970. **MR 42**#3552.

[Fr]   FROSTMAN, O.: *Potential d'equilibre et capacité des ensembles avec quelques applications à la théorie des fonctions* Meddel. Lunds Univ. Mat. Sem. **3** (1935), 1 - 118.

[Fuhr]   FUHRMANN, P.A.: *On Hankel operator ranges, meromorphic pseudo-continuations and factorization of operator-valued analytic functions.* J. Lond. Math. Soc. (2), **13** (1976), 323 - 327. **MR 55**#1128.

[Gar]   GARNETT, J.B.: *Bounded Analytic Functions.* New York and London: Academic Press 1981. **MR 83g**:30037.

[Gin]   GINZBURG, Ju.P.: *On multiplicative representations of J-nonexpansive operatorfunctions* (in Russian). *Part* I. Mat. Issled. **2** (1967), no.2, 52 - 83; *Part* II. Mat. Issled. **2** (1967), no.3, 20 - 51. Engl. transl. in: Amer. Math. Soc. Transl. (**2**), **96** (1970), 189 - 254. **MR 38**#1551.

[Gon]   GONČAR, A.A.: *On a generalized analytic continuation* (in Russian). Math. Sbornik, **76** (1968), no.1, 135 - 146. Engl. transl. in: Math. USSR Sbornik **5** (1968), no.1, 129 - 140. **MR 38**#323.

[Gr-Nik] GRIBOV, M.B., and N.K. NIKOL'SKIĬ: *Invariant subspaces and rational approximation* (in Russian). Zap. Naučn. Sem. Leningrad. Otdel. Mat. Inst. Steklov (LOMI) **92** (1979), 103 - 114, 320. **MR 81f**:47006.

[Hi-Wa]   HILDEN, H.M., and L.J. WALLEN: *Some cyclic and noncyclic vectors of certain operators.* Indiana Univ. Math. J. **23** (1973/74), 557 - 565. **MR 48**#4796.

[Hof]   HOFFMANN, K.: *Banach Spaces of Analytic Functions.* Englewood Cliffs, N.J.: Prentice-Hall 1962. **MR 24**#A2844.

[Kats1]   KATSNELSON, V.: *Weight approximation of entire functions of exponential type of several variables by trigonometrical polynomials* (in Russian). Zap. Naučn. Sem. Leningrad. Otdel. Mat. Inst. Steklov (LOMI) **65** (1976), 69 - 79. Engl. transl. in: Journal of Soviet Math. **16** (1981), 1095 - 1101. **MR 58**#6309.

[Kats2]   KATSNELSON, V.: *Weight spaces of pseudocontinuable functions and approximations by rational functions with prescribed poles.* Zeitschr. für Analysis und ihre Anwend. **12** (1993), 27 - 67.

[Koo1]   KOOSIS, P.: *Sur l'approximation pondérée par des polynomes et par des sommes
           d'exponentielles imaginaires* Ann. Sci. École Norm. Sup. (3) **81** (1964), 387 - 408.
           MR 31#1505.

[Koo2]   KOOSIS, P.: *Introduction to H_p Spaces.* Cambridge: Cambridge University Press 1980.
           MR 81c:30062.

[Kr]   KRIETE III, T.L.: *A generalized Paley-Wiener theorem.* J. Math. Anal. and Appl. **36**
           (1971), 529 - 555. **MR 44**#5473.

[Lev]   LEVIN, M.B.: *Estimation of the derivative of a meromorphic function on the boundary
           of the domain* (in Russian). Teor. Funkcĭ Funkcional. Anal. i Priložen **24** (1975),
           (Marchenko V.A. - ed.), 68 - 75. **MR 53**#5876.

[L-McK]   LEVINSON, N., and H.P. MCKEAN: *Weight trigonometrical approximation on R^1 with
           application to the germ field of a stationary Gaussian noise.* Acta Math. **112** (1964),
           99 - 143. **MR 29**#414.

[Lju]   LJUBARSKIĬ, Ju.: *The approximation of entire functions of exponential type by trigono-
           metrical polynomials* (in Russian). Mat. Fiz. i Funkcional. Anal. **3** (1972), (Marchenko
           V.A. - ed.), 56 - 67, 108. **MR 57**#10335.

[Mar]   MARSHALL, D.E.: *Blaschke products generate H^∞.* Bull. Amer. Math. Soc. **82**
           (1976), 494 - 496. **MR 53**#5877.

[Ne]   NEVANLINNA, R.:   *Über eine Klasse meromorpher Functionen.*   Math. Ann. **92**
           (1924), 145 -154.

[Ne-Ne]   NEVANLINNA, F., and R. NEVANLINNA: *Über die Eigenschaften analytischer Funktionen
           in der Umgebung einer singulären Stelle oder Linie.* Acta. Soc. Sci. Fenn. **50** (1922),
           no.5.

[Nik1]   NIKOL'SKIĬ, N.K.: *Lectures on the shift operator* (in Russian). Zap. Naučn. Sem.
           Leningrad. Otdel. Mat. Inst. Steklov (LOMI) **39** (1974), 59 - 93. Engl. transl. in:
           Journal of Soviet Math. **8** (1977), 41 - 64. **MR 50**#1025.

[Nik2]   NIKOL'SKIĬ, N.K.: *Treatise on the Shift Operator.*   Berlin-Heidelberg-New York:
           Springer-Verlag 1986. **MR 87i**:47042.

[Ni-Kh]   NIKOL'SKIĬ, N.K. and V. P. KHAVIN: *The results of V.I. Smirnov in complex analysis
           and their further development* (in Russian). In: SMIRNOV, V.I.: Selected Papers - Com-
           plex Analysis. Mathematical Theory of Diffraction (in Russian). Leningrad: Leningrad
           University Press 1988, pp. 111 - 145. **MR 91g**:01040.

[O]   OSTROWSKIĬ, A.: *Über vollständige Gebiete gleimässigev Konvergenz von Folgen ana-
           lytischer Funktionen.* Abh. Math. Sem. Hamburg **1** (1922), 327 - 350. Reprinted in:
           OSTROWSKIĬ, A.: Collected Mathematical Papers. Vol. 5 (Complex Function Theory).
           Basel: Birkhäuser 1984, pp. 22 - 45.

[Pot1]   POTAPOV, V.P.: *On holomorphic matrix functions bounded in the unit disc* (in Rus-
           sian). Engl. transl. in: Collected Papers of V.P. Potapov. Sapporo. Japan: Private
           translation and edition by T. Ando 1982, pp. 7 - 12. **MR 13**, 536 - 537.

[Pot2]   POTAPOV, V.P.:   *The multiplicative structure of J-contractive matrix functions* (in Russian). Engl. transl. in: Amer. Math. Soc. Transl. (2), **15** (1960), 131 - 243. **MR 17**, 958 - 959.

[RoRo1] ROSENBLUM, M., and J. ROVNYAK: *The factorization problem for nonnegative operator-valued functions.* Bull. Amer. Math. Soc. **77** (1971), 287 - 318. **MR 42#8315**.

[RoRo2] ROSENBLUM, M., and J. ROVNYAK: *Cayley Inner Functions and Best Approximation.* Journ. of Approxim. Theory. **17** (1976), 241 - 253. **MR 58#29632**.

[RoRo3] ROSENBLUM, M., and J. ROVNYAK: *Change of variables formulas with Cayley inner functions.* In: Topics in Functional Analysis (Essays dedicated to M.G.Krein on the occasion of his 70th birthday). Advances in Math., Supl. Studies **3**. New York and London: Academic Press 1978, pp. 283 - 320. **MR 81d:30053**.

[Rud1]   RUDIN, W.: *A generalization of a theorem of Frostman.* Math. Scand. **21** (1967), 136 - 143. **MR 38#3463**.

[Rud2]   RUDIN, W.: *Function Theory in Polydiscs.* New York-Amsterdam: Benjamin 1969. **MR 41#501**.

[Sh1]   SHAPIRO, H.S.: *Weighted polynomial approximation and boundary behavior of analytic functions.* In: Contemporary Problems in Theory Anal. Functions (Internat. Conf., Erevan, 1965). Moscow 1966, pp. 326 -335. **MR 35#383**.

[Sh2]   SHAPIRO, H.S.:   *Generalized analytic continuation.* In: Symposia on Theoretical Physics and Mathematics, Vol. 8 (Symposium, Madras, 1967). New York: Plenum 1968, pp. 151 - 163. **MR 39#2953**.

[Sh3]   SHAPIRO, H.S.: *Overconvergence of sequences of rational functions with sparse poles.* Ark. Math. **7** (1968), 343 - 349. **MR 38#4658**.

[Sm]   SMIRNOV, V.I.: *Sur les formules de Cauchy et de Green et quelques problèmes qui s'y rattachent* (in French). Izv. Akad. Nauk SSSR, Ser. Mat. **3** (1932), 338 - 372. Russian transl. in: SMIRNOV, V.I.: Selected Papers - Complex Analysis. Mathematical Theory of Diffraction (in Russian). Leningrad: Leningrad University Press 1988, pp. 82 - 111. **MR 91g:01040**.

[So-Yu] SODIN, M. L., and P. M. YUDITSKIĬ: *Infinite-zone Jakobi matrices with pseudocontinuable Weyl's functions and a homogeneous spectrum* (in Russian). To appear in: Dokl. Akad. Nauk (Rossiĭskaya Akad. Nauk) (1993).

[Sz]   SZEGÖ, G.:   *Über die Randwerte einer analytischen Funktionen* . Math. Ann. **84** (1961), 232 - 244. Reprinted in: SZEGÖ, G.: Collected Papers. Vol. 1. Basel: Birkhäuser 1982, pp. 404 - 416.

[Tum1]   TUMARKIN, G.Ts.: *Sequences of Blaschke products* (in Russian). Dokl. Akad. Nauk SSSR **129** (1959), 40 - 43. **MR 21#7308**.

[Tum2]   TUMARKIN, G.Ts.: *The decomposition of analytic functions in series of rational fractions with a given set of poles* (in Russian). Izv. Akad. Nauk Armjan. SSR. Ser. Fiz-Mat. Nauk **14** (1961), no.1, 9 - 31. **MR 26#333**.

[Tum3]   TUMARKIN, G.Ts.: *Conditions for the existence of an analytic majorant for a family of analytic functions* (in Russian). Izv. Akad. Nauk Armjan. SSR. Ser. Fiz-Mat. Nauk **17** (1964), no.6, 3 - 25. **MR 30**#3976.

[Tum4]   TUMARKIN, G.Ts.: *Convergent sequences of Blaschke products* (in Russian). Sibirsk. Mat. Ž. **5** (1964), 201 - 233. **MR 28**#5182.

[Tum5]   TUMARKIN, G.Ts.: *Approximation with respect to various metrics of functions defined on the circumference by sequences of rational functions with fixed poles* (in Russian). Izv. Akad. Nauk SSSR. Ser. Mat. **30** (1966), 721 - 766. Engl. transl. in: Amer. Math. Soc. Transl. (**2**), **77** (1968), 183 - 233. **MR 33**#7562.

[Tum6]   TUMARKIN, G.Ts.: *Necessary and sufficient conditions for the possibility of approximating a function on a circumference by rational functions, expressed in terms directly connected with the distribution of poles of the approximating fractions* (in Russian). Izv. Akad. Nauk SSSR. Ser. Mat. **30** (1966), 969 - 980. Engl. transl. in: Amer. Math. Soc. Transl. (**2**), **77** (1968), 235 - 248. **MR 34**#2905.

[Tum7]   TUMARKIN, G.Ts.: *Description of a class of functions admitting an approximation by fractions with preassigned poles* (in Russian). Izv. Akad. Nauk Armjan. SSR. Ser. Mat. **1** (1966), no.2, 85 - 109. **MR 34**#6123.

[Tum8]   TUMARKIN, G.Ts.: *Conditions for the convergence of boundary values of analytic functions and approximation of rectifiable curves* (in Russian). In: Contemporary Problems in Theory Anal. Functions (Internat. Conf., Erevan, 1965). Moscow 1966, 283 - 295. **MR 34**#7808.

[Tum9]   TUMARKIN, G.Ts.: *Boundary properties of sequences of analytic functions* (in Russian). Thesis for a Doctor's degree. Leningrad 1961.

[VP]   DE LA VALEÉ POUSSIN: *Balayage et Représentation Conforme.* Paris: Gauthier-Villars 1949.

[Vin]   VINOGRADOV, S.A.: *Properties of multipliers of Cauchy-Stieltjes integrals and some factorization problems for analytic functions* (in Russian). In: Mathematical Programming and Related Questions (Proc. Seventh Winter School, Drogobych, 1974: Theory of Functions and Functional Analysis). Moscow: Central Econom.-Math. Inst. Akad. Nauk SSSR 1976, pp. 5 - 39. Engl. Transl. in: Amer. Soc. Transl. (**2**) **115** (1980) 1 - 32. **MR 58**#28518.

[W]   WALSH, J.L.: *On interpolation and approximation by rational functions with preassigned poles.* Trans. Amer. Math. Soc. **34** (1932), 22 - 74.

Department of Theoretical Mathematics
The Weizmann Institute of Science   E-mail: katze@wisdom.weizmann.ac.il
Rehovot 76100, Israel

AMS subject classification: 30D10, 30D15, 46B15, 94A12.

Operator Theory:
Advances and Applications, Vol. 72
© 1994 Birkhäuser Verlag Basel

AN ANALYSIS AND EXTENSION OF V.P. POTAPOV'S APPROACH TO INTERPOLATION PROBLEMS WITH APPLICATIONS TO THE GENERALIZED BI-TANGENTIAL SCHUR-NEVANLINNA-PICK PROBLEM AND J-INNER-OUTER FACTORIZATION

A. Ya. Kheifets and P. M. Yuditskii

There exist several approaches to the generalization and unification of various problems of the Nevanlinna-Pick type [1–4, 5, 9, 10, 12, 14, 28, 29, 31, 35, 38, 40].

V.P. Potapov was the founder of one of such approach [13, 27]. The key element in the method he proposed is the reduction of the interpolation problem to the solution of an inequality that the unknown functions must satisfy. He named this inequality the Fundamental Matrix Inequality (FMI). The solution of the FMI is based on a special kind of identity called the Fundamental Identity (FI).

Potapov's approach was developed by a number of mathematicians. The work in this field led the authors, together with V.E. Katsnelson, to the setting and solution of an Abstract Interpolation Problem (AIP). While doing this, they discovered a close link between Potapov's approach to interpolation problems and the approach based on the extension theory of operators [1–4, 8, 30].

This paper is organized as follows. An outline of the approach initiated by Potapov is given in Section 1 for the Nevanlinna-Pick problem, which serves as a basic example. In Section 2 we introduce some ideas which lead to the Abstract FI and the Abstract FMI, and also describe the FMI transformation (Theorem 2.1). In Section 3 the setting of the AIP is given as a consequence of the analysis of the transformed FMI. The link between the AIP and the theory of extensions of isometric operators is explained in Section 4. This is the basis for the AIP solution (the Main Theorem of Section 4). Section 5 deals with the application of the AIP scheme to the generalized bi-tangential Schur-Nevanlinna-Pick problem. In particular, the generalization of the well-known Nevanlinna and Adamjan-Arov-Krein theorem on unimodular solutions is presented here (Theorems 5.10, 5.11). This generalization concerns the so-called semi-determinate case of solvability. In Section 6, the J-inner-outer factorization theorem is presented as another application of the AIP. Its proof is based on the construction of a J-outer factor from the boundary values of the J-form of a factorable matrix-function.

We would like to mention that the list of references does not claim to be complete. It includes only the references that were used for this paper.

We wish to thank L.A. Sahnovich who kindly gave us an opportunity to participate in this collection, M. Sodin for helpful discussions, and E. Romanovskaja for technical assistance.

1. POTAPOV'S APPROACH TO THE NEVANLINNA-PICK PROBLEM

The first problem to which Potapov applied his method [27] was the Nevanlinna-Pick interpolation problem:

Let $\zeta_1, \ldots, \zeta_n$ be points of the open unit disk $\mathbb{D}$, and let $\omega_1, \ldots, \omega_n$ be complex numbers.

It is required to describe all functions ω which are both holomorphic and contractive (i.e., $|\omega(\zeta)| \le 1$,) in $\mathbb{D}$, such that $\omega(\zeta_k) = \omega_k$.

In what follows, we denote the class of functions which are both holomorphic and contractive in $\mathbb{D}$ by $\mathbb{B}$.

The Nevanlinna-Pick problem is not solvable for certain interpolation data. The following lemma gives necessary and sufficient conditions for its solvability.

SCHWARZ-PICK LEMMA. *Let* $\omega \in \mathbb{B}$, $\omega(\zeta_k) = \omega_k$. *Then the quadratic form generated by the matrix*

$$
D = \begin{bmatrix}
\dfrac{1 - \bar{\omega}_1 \omega_1}{1 - \bar{\zeta}_1 \zeta_1} & \cdots & \dfrac{1 - \bar{\omega}_1 \omega_n}{1 - \bar{\zeta}_1 \zeta_n} \\
\cdots & \cdots & \cdots \\
\dfrac{1 - \bar{\omega}_n \omega_1}{1 - \bar{\zeta}_n \zeta_1} & \cdots & \dfrac{1 - \bar{\omega}_n \omega_n}{1 - \bar{\zeta}_n \zeta_n}
\end{bmatrix}
\tag{1.1}
$$

is nonnegative: $\langle Dx, x \rangle \ge 0$, $x \in \mathbb{C}^n$.

This is the starting point of Potapov's investigation of the Nevanlinna-Pick problem. Adding an arbitrary point $\zeta \in \mathbb{D}$ to the set of points $\zeta_1, \ldots, \zeta_n$ and a value

$\omega = \omega(\zeta)$ to the set of values $\omega_1, \ldots, \omega_n$, Potapov applied the Schwarz-Pick Lemma to the expanded data:

$$
\begin{bmatrix}
& & & \Big| & \dfrac{1 - \overline{\omega}_1 \omega(\zeta)}{1 - \overline{\zeta}_1 \zeta} \\
& & & \Big| & \\
& D & & \Big| & \vdots \\
& & & \Big| & \\
& & & \Big| & \dfrac{1 - \overline{\omega}_n \omega(\zeta)}{1 - \overline{\zeta}_n \zeta} \\
\hline
\dfrac{1 - \overline{\omega(\zeta)} \omega_1}{1 - \overline{\zeta} \zeta_1} & \cdots & \dfrac{1 - \overline{\omega(\zeta)} \omega_n}{1 - \overline{\zeta} \zeta_n} & \Big| & \dfrac{1 - \overline{\omega(\zeta)} \omega(\zeta)}{1 - \overline{\zeta} \zeta}
\end{bmatrix}
\geq 0 . \qquad (1.2)
$$

This inequality is a constraint on the unknown function $\omega(\zeta)$. All other values in this inequality are given. Potapov's idea is to find solutions of the interpolation problem as solutions of inequality (1.2). He called it the Fundamental Matrix Inequality (FMI) of the Nevanlinna-Pick problem.

As a preliminary step, he proved that (1.2) has no "unnecessary" solutions that are not solutions of the interpolation problem:

THEOREM. *A function $\omega(\zeta)$ which is holomorphic in* $\mathbb{D}$ *solves the Nevanlinna-Pick problem iff it satisfies the inequality (1.2) for every $\zeta \in \mathbb{D}$.*

To investigate the FMI, Potapov rewrote it making use of a matrix generalization of the Sylvester criterion which he called the Block-Matrix Lemma: *Let $D > 0$, then (1.2) is equivalent to the inequality*

$$
\frac{1 - \overline{\omega(\zeta)} \omega(\zeta)}{1 - \overline{\zeta} \zeta} - \begin{bmatrix} \dfrac{1 - \overline{\omega(\zeta)} \omega_1}{1 - \overline{\zeta} \zeta_1} & \cdots & \dfrac{1 - \overline{\omega(\zeta)} \omega_n}{1 - \overline{\zeta} \zeta_n} \end{bmatrix} D^{-1} \begin{bmatrix} \dfrac{1 - \overline{\omega}_1 \omega(\zeta)}{1 - \overline{\zeta}_1 \zeta} \\ \vdots \\ \dfrac{1 - \overline{\omega}_n \omega(\zeta)}{1 - \overline{\zeta}_n \zeta} \end{bmatrix} \geq 0 . \qquad (1.3)
$$

The left-hand side of (1.3) is a quadratic form in the unknown function ω. The coefficients of this form depend only on the interpolation data and the independent variable $\zeta \in \mathbb{D}$, i.e., (1.3) may be rewritten in the following form

$$
\begin{bmatrix} \overline{\omega(\zeta)}, 1 \end{bmatrix} \Phi(\zeta, \overline{\zeta}) \begin{bmatrix} \omega(\zeta) \\ 1 \end{bmatrix} \geq 0 , \qquad (1.4)
$$

where

$$\Phi(\zeta,\overline{\zeta}) = j - (1-\overline{\zeta}\zeta) \begin{bmatrix} \dfrac{-\omega_1}{1-\overline{\zeta}\zeta_1} & \cdots & \dfrac{-\omega_n}{1-\overline{\zeta}\zeta_n} \\[2mm] \dfrac{1}{1-\overline{\zeta}\zeta_1} & \cdots & \dfrac{1}{1-\overline{\zeta}\zeta_n} \end{bmatrix} D^{-1} \begin{bmatrix} \dfrac{-\overline{\omega}_1}{1-\overline{\zeta}_1\zeta} & \dfrac{1}{1-\overline{\zeta}_1\zeta} \\[2mm] \vdots & \vdots \\[2mm] \dfrac{-\overline{\omega}_n}{1-\overline{\zeta}_n\zeta} & \dfrac{1}{1-\overline{\zeta}_n\zeta} \end{bmatrix}$$

and

$$j = \begin{bmatrix} -1 & 0 \\ 0 & 1 \end{bmatrix}.$$

To obtain the representation for the function $w(\zeta)$ from (1.4), Potapov expressed the 2×2 matrix function $\Phi(\zeta,\overline{\zeta})$ in the form

$$\Phi(\zeta,\overline{\zeta}) = \mathcal{U}(\zeta)^{-1*} j \mathcal{U}(\zeta)^{-1}, \tag{1.5}$$

where $\mathcal{U}(\zeta)$ is a meromorphic 2×2 matrix function. Then the inequality (1.4) takes the form

$$[\overline{w(\zeta)},1]\mathcal{U}(\zeta)^{-1*} j \mathcal{U}(\zeta)^{-1} \begin{bmatrix} w(\zeta) \\ 1 \end{bmatrix} \geq 0. \tag{1.6}$$

The last inequality means that for the pair

$$\begin{bmatrix} p \\ q \end{bmatrix} = \mathcal{U}^{-1} \begin{bmatrix} w \\ 1 \end{bmatrix} \tag{1.7}$$

the inequality

$$[\overline{p},\overline{q}]j \begin{bmatrix} p \\ q \end{bmatrix} \geq 0 \tag{1.8}$$

holds. Thus the function $\varepsilon = pq^{-1}$ belongs to the class $\mathbb{B}$. By (1.7)

$$\begin{bmatrix} w \\ 1 \end{bmatrix} = \mathcal{U} \begin{bmatrix} p \\ q \end{bmatrix},$$

whence

$$w = (\alpha_{11}\varepsilon + \alpha_{12})(\alpha_{21}\varepsilon + \alpha_{22})^{-1} \tag{1.9}$$

where the $\alpha_{i,j}$ are the entries of $\mathcal{U}$:

$$\mathcal{U} = \begin{bmatrix} \alpha_{11} & \alpha_{12} \\ \alpha_{21} & \alpha_{22} \end{bmatrix}.$$

Thus every solution of the inequality (1.2) has the form (1.9) where $\mathcal{U}$ is uniquely determined from (1.5) (up to a left j-unitary constant factor) and ε is any function of the class $\mathbb{B}$.

Conversely, upon defining a function ω by an arbitrary function $\varepsilon \in \mathbb{B}$ according to formula (1.9) and reversing the arguments, it can be shown that $\omega(\zeta)$ satisfies (1.2), i.e., (1.9) *gives the general form of a solution to the Nevanlinna-Pick Problem.*

Let us mention some properties of the matrix $\mathcal{U}(\zeta)$:

1. $j - \mathcal{U}(\zeta)^* j \mathcal{U}(\zeta) \leq 0 \qquad (|\zeta| < 1),$

2. $j - \mathcal{U}(\zeta)^* j \mathcal{U}(\zeta) = 0 \qquad (|\zeta| = 1),$

3. $\mathcal{U}(\zeta)$ is a Blaschke-Potapov product with poles at the points $\zeta_1, \ldots, \zeta_n$.

We mention that the principal motivation which stimulated Potapov's interest in classical interpolation problems was his investigation of the structure and parametrization of j-contractive meromorphic matrix-functions by interpolation data [13, 26, 27].

2. AN ANALYSIS OF POTAPOV'S APPROACH AND THE AIP

We begin the analysis of Potapov's approach with the following remark. In the previous section it was not explained how one can obtain the representation (1.5). This representation was essential for the given construction. A careful analysis of its proof shows that the representation (1.5) can be obtained by purely algebraic means. This proof is based on the following identity

$$ D - \begin{bmatrix} \overline{\zeta_1} & & \\ & \ddots & \\ & & \overline{\zeta_n} \end{bmatrix} D \begin{bmatrix} \zeta_1 & & \\ & \ddots & \\ & & \zeta_n \end{bmatrix} = \begin{bmatrix} 1 \\ \vdots \\ 1 \end{bmatrix} [1 \cdots 1] - \begin{bmatrix} \overline{\omega}_1 \\ \vdots \\ \overline{\omega}_n \end{bmatrix} [\omega_1, \ldots, \omega_n] . \quad (2.1) $$

Identity (2.1) is called the Fundamental Identity (FI) and it forms the basis of our subsequent constructions.

Let X denote the n-dimensional complex space $\mathbb{C}^n$; the vectors of this space are columns.

We have already denoted the matrix in (1.1) by D. In what follows we will use the same symbol to denote the quadratic form

$$ D(x, x) = \langle Dx, x \rangle $$

on X, where $\langle x, y \rangle$ is the usual inner product in $\mathbb{C}^n$. We let T denote the linear operator on X which is defined by the matrix

$$T = \begin{bmatrix} \zeta_1 & & \\ & \ddots & \\ & & \zeta_n \end{bmatrix}$$

and let M_1 and M_2 denote the linear functionals on X which are defined by the formulas

$$M_1 x = [1, \ldots, 1] x \ ,$$

$$M_2 x = [\omega_1, \ldots, \omega_n] x \ . \tag{2.2}$$

Then the identity (2.1) may be rewritten in the form

$$D(x, y) - D(Tx, Ty) = \overline{M_1 y} \cdot M_1 x - \overline{M_2 y} \cdot M_2 x \ , \tag{2.3}$$

where x and y are arbitrary vectors from X. In terms of the notation which has been introduced above, inequality (1.2) has the form

$$\left[\begin{array}{c|c} D(x, x) & \overline{(M_1 - \overline{\omega(\zeta)}M_2)(I - \bar{\zeta}T)^{-1}x} \\ \hline (M_1 - \overline{\omega(\zeta)}M_2)(I - \bar{\zeta}T)^{-1}x & \dfrac{1 - \overline{\omega(\zeta)}\omega(z)}{1 - \bar{\zeta}\zeta} \end{array} \right] \geq 0 \ , \tag{2.4}$$

for arbitrary $x \in X$.

Using the identity (2.3) one can rewrite inequality (2.4) in another form which is equivalent to (2.4):

$$\left[\begin{array}{c|c} D(x, x) & \overline{(\omega(\zeta)M_1 - M_2)(\zeta - T)^{-1}x} \\ \hline (\omega(\zeta)M_1 - M_2)(\zeta - T)^{-1}x & \dfrac{1 - \omega(\zeta)\overline{\omega(z)}}{1 - \bar{\zeta}\zeta} \end{array} \right] \geq 0 \ , \tag{2.5}$$

Now we drop the specific choices of the operators D, T and the functionals M_1, M_2 on $X = \mathbb{C}^n$ which were made in the last section, (here we follow A. A. Nudelman

[35, 37]) and claim that: *Potapov's method as explained in Section 1 is applicable to the solution of inequality (2.4) or (2.5), once identity (2.3) holds and D > 0 [26].* A description of the set of solutions of this inequality by a formula like (1.9) is a result of the present procedure. We will call these inequalities the FMI, and identity (2.3) the FI as before.

A question arises: What is the meaning of the FMI as far as interpolation is concerned? A partial answer to this question is rather simple. Inequality (2.5) implies that the function

$$(F_+x)(\zeta) = (\omega(\zeta)M_1 - M_2)(\zeta I - T)^{-1}x \qquad (2.6)$$

has removable singularities in $\mathbb{D}$, i.e., it is holomorphic for every $x \in X$. It also follows from (2.4) that the function

$$(F_-x)(\zeta) = \overline{\zeta}(M_1 - \overline{\omega(\zeta)}M_2)(1 - \overline{\zeta}T)^{-1}x \qquad (2.7)$$

is anti-holomorphic in $\mathbb{D}$ for every $x \in X$.

EXAMPLE. For the Nevanlinna-Pick problem

$$(F_+x)(\zeta) = \sum_{k=1}^{n} \frac{\omega(\zeta) - \omega_k}{\zeta - \zeta_k} x_k \; .$$

The analyticity of this function for arbitrary vectors $\{x_k\} \in \mathbb{C}^n$ means precisely that $\omega(\zeta_k) = \omega_k$.

We face a more difficult situation when T has spectral points on the circle $\mathbb{T}$. Not all the analytical restrictions on the function $\omega(\zeta)$ are immediately obvious from the FMI in this case. I.V. Kovalishina met this obstacle while investigating the power moment problem (there T is a Jordan cell with an eigenvalue on $\mathbb{T}$). She transformed the FMI for the moment problem into a form which made clear the concealed interpolation condition contained in the FMI [26]. It constituted the beginning of the FMI transformation technique which is important in what follows.

Above we dealt with the finite-dimensional space $X = \mathbb{C}^n$. But in numerous problems we encounter infinite-dimensional spaces X (for example, in the continuation problem of M.G. Krein). Potapov's approach is applicable to problems with this kind of space X also, see [17, 25, 41]. Quadratic forms $D \geq 0$, linear operators T and direction functionals M_1, M_2 are present in all of these problems. Their solution is based on the identity (2.3). Therefore it is only natural to introduce the concept of an abstract FMI, i.e., an inequality like (2.4) and an abstract FI like (2.3), see [15, 16, 18, 19].

We now return to the question of the interpolation sense of the *abstract* FMI. Transformation techniques have been developed to a considerable extent in papers by V.E. Katsnelson [27], T.S. Ivanchenko and L.A. Sahnovich [15, 16]. They form the basis for answering this question. We present here the corresponding theorem from [20].

THEOREM 2.1. *ω is a solution of the FMI (2.4) iff*

$$
\left[
\begin{array}{c|cc}
P_\zeta(x,x) + \overline{P_\zeta(x,x)} & \overline{(F_+x)(\zeta)} & \overline{(F_-x)(\zeta)} \\
\hline
(F_+x)(\zeta) & 1 & \omega(\zeta) \\
(F_-x)(\zeta) & \overline{\omega(\zeta)} & 1
\end{array}
\right] \geq 0 , \quad (2.8)
$$

for every $x \in X$, where $(F_+x)(\zeta)$, $(F_-x)(\zeta)$ are defined by formulas (2.6), (2.7) and

$$
P_\zeta(x,y) = \frac{1}{2}D\left(x, \frac{1+\overline{\zeta}T}{1-\overline{\zeta}T}y\right) - \zeta\overline{M_2(I - \overline{\zeta}T)^{-1}y} \cdot (F_+x)(\zeta) .
$$

Note that $P_0(x,y) = \frac{1}{2}D(x,y)$.

The inequality (2.8) is called the Transformed FMI (TFMI). In what follows we will use the following notation:

$$
(Fx)(\zeta) = \begin{bmatrix} (F_+x)(\zeta) \\ (F_-x)(\zeta) \end{bmatrix} . \tag{2.9}
$$

The TFMI (2.8) allows us to obtain full information about the analytic conditions which are contained in the abstract FMI [20].

THEOREM 2.2. *A function $\omega(\zeta) \in \mathbb{B}$ satisfies inequality (2.8) iff the following conditions are fulfilled for every $x \in X$:*

1. $F_+x \in H_+^2$, $F_-x \in H_-^2$, $(H_+^2, H_-^2$ *are the usual Hardy spaces)*,

2. $(Fx)(t) \in \text{Ran} \begin{bmatrix} 1 & \omega(t) \\ \overline{\omega}(t) & 1 \end{bmatrix}$, *for a.e. $t \in \mathbb{T}$,* $\qquad (2.10)$

3. $\int\langle \begin{bmatrix} 1 & \omega \\ \overline{\omega} & 1 \end{bmatrix}^{[-1]} Fx, Fx\rangle dm \leq D(x,x)$,

where dm is normalized Lebesgue measure on $\mathbb{T}$, and the notation $\begin{bmatrix} 1 & \omega \\ \overline{\omega} & 1 \end{bmatrix}^{[-1]} Fx$ denotes an arbitrary pre-image (clearly the result does not depend on this choice).

Theorem 2.2 is the main motivation for posing the Abstract Interpolation Problem (AIP). But we make some remarks before presenting it.

REMARKS.

1) One can associate with an arbitrary function $\omega \in \mathbb{B}$ the functional space H_ω. It is the model space for a contractive operator with ω as its characteristic function [32, 34, 40]. Here we present it in the de Branges-Rovnyak form [11].

The space H_ω consists of the vector-functions

$$f = \begin{bmatrix} f_+ \\ f_- \end{bmatrix}, \quad f_+ \in H_+^2, \quad f_- \in H_-^2,$$

such that

1. $f(t) \in \text{Ran} \begin{bmatrix} 1 & \omega(t) \\ \omega(t) & 1 \end{bmatrix}$, for a.e. $t \in \mathbb{T}$,

(2.11)

2. $\int \langle \begin{bmatrix} 1 & \omega \\ \bar{\omega} & 1 \end{bmatrix}^{[-1]} f, f \rangle dm < \infty$.

H_ω is a Hilbert space. The integral (2.11) defines the norm on H_ω.

We can rewrite conditions (2.10) from Theorem 2.2 by using this definition:

1. $Fx \in H_\omega$,

2. $\|Fx\|_{H_\omega}^2 \leq D(x, x)$,

(2.12)

for every $x \in X$.

2) It may be easily deduced from the definition (2.6) that

$$[F_+(\zeta I - T)x](\zeta) = (\omega(\zeta)M_1 - M_2)x$$

i.e.,

$$(F_+Tx)(\zeta) = \zeta(F_+x)(\zeta) - (\omega(\zeta)M_1 - M_2)x \qquad (2.13)$$

and similarly that

$$(F_-Tx)(\zeta) = \frac{1}{\zeta}(F_-x)(\zeta) - (M_1 - \overline{\omega(\zeta)}M_2)x . \qquad (2.14)$$

Letting the independent variable ζ tend to the boundary of $\mathbb{D}$ ($|\zeta| \to 1$) and combining (2.13) and (2.14), we get

$$FTx = t \cdot Fx - \begin{bmatrix} 1 & \omega \\ \bar{\omega} & 1 \end{bmatrix} \begin{bmatrix} -M_2x \\ M_1x \end{bmatrix}, \quad \text{for a.e. } t \in \mathbb{T}. \qquad (2.15)$$

If the map $F : X \to H_\omega$ satisfies the property (2.15), and if all the expressions in (2.6) and (2.7) make sense, then F_+ and F_- coincide with (2.6) and (2.7), respectively. Thus we can reformulate Theorem 2.2 in the following way [20]:

THEOREM 2.2'. *A function $\omega(\zeta) \in \mathbb{B}$, satisfies the inequality (2.8) iff there exists a map $F : X \to H_\omega$, which satisfies equality (2.15) and the inequality $\|Fx\|^2_{H_\omega} \leq D(x, x)$.*

Property (2.15) is meaningful even when (2.6) and (2.7) are not. This allows us to choose an operator T in (2.15) without additional spectral restrictions [16, 18]. Note that, in general, F is not defined uniquely by ω from (2.15).

3) Next, let $T = T_1 T_2^{-1}$ where T_2 is invertible, and introduce the new notations $M_i := M_i T_2$, $i = 1, 2$, so that equality (2.15) takes the form

$$FT_1 x = t \cdot FT_2 x - \begin{bmatrix} 1 & \omega \\ \overline{\omega} & 1 \end{bmatrix} \begin{bmatrix} -M_2 x \\ M_1 x \end{bmatrix} , \quad \text{for a.e. } t \in \mathbf{T} .$$

This is clearly meaningful even if T_2 is not invertible.

In the following section we pose the AIP; it is a natural generalization of the problem of describing the FMI solutions set.

3.   THE ABSTRACT INTERPOLATION PROBLEM

In the previous section we considered problems dealing with a scalar function $\omega \in \mathbb{B}$. Potapov investigated and attached great importance to problems dealing with matrix-valued functions. We pose the Abstract Interpolation Problem for operator-valued functions.

First we would like to recall the definition of the de Branges-Rovnyak space [11, 34] associated with an arbitrary operator-function (o.f. for short) $\omega \in \mathbb{B}[E_1, E_2]$, where E_1, E_2 are separable Hilbert spaces, $[E_1, E_2]$ is the space of bounded linear operators from E_1 to E_2, $\mathbb{B}[E_1, E_2]$ is the set of $[E_1, E_2]$-valued operator-functions which are holomorphic and contractive on $\mathbb{D}$.

Let $L^2(E)$, $H^2_+(E)$, $H^2_-(E)$ be the standard notations for E-valued vector-function spaces [33, 40], and let L^2_ω be the space of vector-valued functions $f(t)$ such that

$$f(t) = \begin{bmatrix} I_{E_2} & \omega(t) \\ \omega(t)^* & I_{E_1} \end{bmatrix}^{\frac{1}{2}} \tilde{f}(t) ,$$

where $\widetilde{f} \in L^2(E_2 \oplus E_1)$ and

$$\widetilde{f}(t) \perp \mathrm{Ker} \begin{bmatrix} I & \omega(t) \\ \omega(t)^* & I \end{bmatrix} \qquad \text{for a.e. } t \in \mathbf{T},$$

(the orthogonality is with respect to the original inner product in $E_2 \oplus E_1$). We will use the following notation for $\widetilde{f}$:

$$\widetilde{f}(t) = \begin{bmatrix} I_{E_2} & \omega(t) \\ \omega(t)^* & I_{E_1} \end{bmatrix}^{[-\frac{1}{2}]} f(t),$$

and define the inner product in L_ω^2 by the formula

$$\langle f, g \rangle_{L_\omega^2} \overset{\text{def}}{=} \langle \begin{bmatrix} I_{E_2} & \omega \\ \omega^* & I_{E_1} \end{bmatrix}^{[-\frac{1}{2}]} f, \begin{bmatrix} I_{E_2} & \omega \\ \omega^* & I_{E_1} \end{bmatrix}^{[-\frac{1}{2}]} g \rangle_{L^2(E_2 \oplus E_1)}.$$

L_ω^2 is a complete Hilbert space with respect to this inner product.

Let $f \in L_\omega^2$ and let $f(t) \in \mathrm{Ran} \begin{bmatrix} I_{E_2} & \omega(t) \\ \omega(t)^* & I_{E_1} \end{bmatrix}$ for almost every $t \in \mathbf{T}$ (these functions are dense in L_ω^2; if $\dim E_1 < \infty$ and $\dim E_2 < \infty$, then every function in L_ω^2 is of this form). Then for any $g \in L_\omega^2$:

$$\langle f, g \rangle_{L_\omega^2} = \int_{\mathbf{T}} \langle \begin{bmatrix} I_{E_2} & \omega(t) \\ \omega(t)^* & I_{E_1} \end{bmatrix}^{[-1]} f(t), g(t) \rangle_{E_2 \oplus E_1} dm(t),$$

where just as before $\begin{bmatrix} I_{E_2} & \omega(t) \\ \omega(t)^* & I_{E_1} \end{bmatrix}^{[-1]} f(t)$ denotes an arbitrary pre-image. Sometimes we use the last form of inner product notation for all $f \in L_\omega^2$ in order to simplify the typography (keeping in mind the definitions presented above).

The de Branges-Rovnyak space H_ω is the subspace of L_ω^2 which consists of the functions

$$H_\omega = \left\{ \begin{bmatrix} f_+ \\ f_- \end{bmatrix} \in L_\omega^2 \mid f_+ \in H_+^2(E_2) \text{ and } f_- \in H_-^2(E_1) \right\}.$$

Now, following [20], we pose:

THE ABSTRACT INTERPOLATION PROBLEM.

Let

X be a complex linear space;

E_1 and E_2 be separable Hilbert spaces;

T_1 and T_2 be linear operators on X;

M_1 and M_2 be linear maps from X to E_1 and E_2 respectively;

D be a nonnegative quadratic form on X;

and suppose that these objects are linked by the following identity:

$$D(T_2x, T_2y) - D(T_1x, T_1y) = \langle M_1x, M_1y \rangle_{E_1} - \langle M_2x, M_2y \rangle_{E_2} , \qquad (3.1)$$

for every $x, y \in X$. It is required to describe the set of all pairs (ω, F) such that

$\omega(\zeta)$ is an $[E_1, E_2]$-valued o.f. which is contractive and analytic on $\mathbb{D}$, and

F is a map from X to the de Branges-Rovnyak space H_ω which possesses the following properties:

$$1. \qquad FT_1x = t \cdot FT_2x - \begin{bmatrix} 1 & \omega \\ \omega^* & 1 \end{bmatrix} \begin{bmatrix} -M_2x \\ M_1x \end{bmatrix} , \quad \textit{for a.e. } t \in \mathbb{T} .$$

$$2. \qquad \|Fx\|^2_{H_\omega} \leq D(x, x) ,$$

(3.2)

for every $x \in X$.

4. THE AIP AND UNITARY EXTENSIONS OF AN ISOMETRY

It appears that the solutions of the AIP are closely connected with the unitary extensions of a certain isometry that may be constructed from the AIP data. This allows us to obtain a formula for the description of all the solutions of the problem.

4.1. The AIP isometry.

The identity (3.1) plays a key role in the construction of the AIP solutions set. It will allow us to connect the isometric operator with the AIP data.

Let us denote by $\bar{x}$ the anti-linear functional on X acting by the rule $\bar{x}(y) = D(x, y)$, $y \in X$, and by $\overline{X}$ the linear space of such functionals. One can introduce an inner product on $\overline{X}$ in the following way: $\langle \bar{x}_1, \bar{x}_2 \rangle = D(x_1, x_2)$, and then complete $\overline{X}$ using this metric (because D is nonnegative). Let H denote the Hilbert space obtained in this way: $H = \text{clos } \overline{X}$.

Now we set

$$d_V = \{\overline{T_1 x} \oplus M_1 x : x \in X\} \subset H \oplus E_1$$

$$\Delta_V = \{\overline{T_2 x} \oplus M_2 x : x \in X) \subset H \oplus E_2 .$$

Then the identity (3.1) means that the correspondence $V : d_V \to \Delta_V$ which is defined by the formula

$$V(\overline{T_1 x} \oplus M_1 x) = \overline{T_2 x} \oplus M_2 x \tag{4.1}$$

is an isometry. We call this operator the AIP isometry. Its defect subspaces are

$$N_{d_V} = H \oplus E_1 \ominus d_V , \quad N_{\Delta_V} = H \oplus E_2 \ominus \Delta_V .$$

4.2. Parametrization of AIP solutions by unitary extensions of an AIP isometry.

The following construction is based on the papers [8, 21]. The space H is called the *state space* of the isometry V and the spaces E_1, E_2 are called the *coefficient spaces*.

Let U be a unitary extension of the isometry V such that only the state space is extended but the coefficient spaces are not. This means that

$$U : K \oplus E_1 \longrightarrow K \oplus E_2 , \quad K \supset H$$

$$U|_{d_V} = V .$$

The space K is called the state space of U.

Each of these extensions generates a solution of the AIP, as we now demonstrate.

DEFINITION 1. *The operator-function*

$$\omega(\zeta) = P_{E_2}(I - \zeta U P_K)^{-1} U|_{E_1} \tag{4.2}$$

is called the scattering operator-function (s.o.f.) of U (with respect to E_1 and E_2).

Note that $\omega \in \mathbb{B}[E_1, E_2]$.

DEFINITION 2. *The map* $\mathbb{F} : K \to H_\omega$

$$\mathbb{F}k = \begin{bmatrix} \mathbb{F}_+ k \\ \mathbb{F}_- k \end{bmatrix} = \begin{bmatrix} P_{E_2}(I - \zeta U P_K)^{-1} U k \\ \zeta P_{E_1}(I - \bar{\zeta} U^* P_K)^{-1} U^* k \end{bmatrix} \tag{4.3}$$

is called the Fourier representation of the space K.

PROPOSITION 4.1. [20, 21]. *The Fourier representation (4.3) satisfies the following properties*

1. $\quad \mathbb{F} P_K U^* + \begin{bmatrix} \omega \\ I_{E_1} \end{bmatrix} P_{E_1} U^* = t \mathbb{F} P_K + \begin{bmatrix} I_{E_2} \\ \omega^* \end{bmatrix} P_{E_2} ,$

$$\text{(4.4)}$$

2. $\quad \|\mathbb{F} k\|_{H_\omega}^2 \leq \|k\|^2 .$

Let us define $F x = \mathbb{F} \overline{x}$. It follows easily from relations (4.4) that F satisfies the properties (3.2), i.e., the pair (ω, F) is a solution of the AIP.

The converse statement is also true: every map F with the properties (3.2) may be obtained in such a way [21].

4.3. Universal extension of the AIP isometry and the parametrization of all AIP solutions.

To describe the set of scattering operator-functions and Fourier representations of unitary extensions (under fixed coefficient spaces) of a given isometry V, it is convenient to use one universal extension of V [8]. This extension is constructed by extending the coefficient spaces but not the state space.

Let N_1, N_2 be copies of the defect spaces N_{d_V}, N_{Δ_V} respectively. Let us adjoin them to coefficient spaces and introduce the operator

$$A : \ H \oplus E_1 \oplus N_2 = d_V \oplus N_{d_V} \oplus N_2 \longrightarrow H \oplus E_2 \oplus N_1 = \Delta_V \oplus N_{\Delta_V} \oplus N_1$$

by the formulas

$$A|_{d_V} = V , \qquad A|_{N_{d_V}} = id : \ N_{d_V} \longrightarrow N_1 . \tag{4.5}$$

$$A|_{N_2} = id : \ N_2 \longrightarrow N_{\Delta_V} ,$$

where id is the identity map.

We will call the s.o.f. of A with respect to the coefficient spaces $E_1 \oplus N_2, E_2 \oplus N_1$,

$$S(\zeta) = P_{E_2 \oplus N_1} (I - \zeta A P_H)^{-1} A|_{E_1 \oplus N_2} , \tag{4.6}$$

the scattering operator function of the AIP, and we will call the corresponding Fourier representation

$$G h = \begin{bmatrix} G_+ h \\ G_- h \end{bmatrix} = \begin{bmatrix} P_{E_2 \oplus N_1} (I - \zeta A P_H)^{-1} A h \\ \overline{\zeta} P_{E_1 \oplus N_2} (I - \overline{\zeta} A^* P_H)^{-1} A^* h \end{bmatrix} , \tag{4.7}$$

the AIP Fourier representation.

MAIN THEOREM. [20, 21]. *The set of all solutions of the AIP is parametrized by the unit ball* $\mathbb{B}[N_1, N_2]$, *where* N_1, N_2 *are the defect spaces of the AIP isometry (4.1).*

For an arbitrary parameter $\omega \in \mathbb{B}[N_1, N_2]$ *the AIP solution* (ω, F) *is defined by the formulas*

$$\omega = P_{E_2}(I_{E_2 \oplus N_1} - S|_{N_2} \varepsilon P_{N_1})^{-1} S|_{E_1} \tag{4.8}$$

$$Fx = \begin{bmatrix} P_{E_2}(I_{E_2 \oplus N_1} - S|_{N_2} \varepsilon P_{N_1})^{-1} G_{+} \bar{x} \\ P_{E_1}(I_{E_1 \oplus N_2} - S^*|_{N_1} \varepsilon^* P_{N_2})^{-1} G_{-} \bar{x} \end{bmatrix}, \tag{4.9}$$

where S *is the AIP s.o.f. (4.6), and* $G : H \to H_S$ *is the AIP Fourier representation (4.7).*

REMARKS.

1) Let us make more precise the meaning of the notation $S|_{N_2} \omega P_{N_1}$. The orthogonal projection P_{N_1} acts from $E_2 \oplus N_1$ onto N_1. Furthermore, $\omega : N_1 \to N_2$ and $S|_{N_2} : N_2 \to E_2 \oplus N_1$. Thus $S|_{N_2} \omega P_{N_1}$ is an operator from $E_2 \oplus N_1$ to $E_2 \oplus N_1$.

2) From the definition of the operator A in (4.5) and $S(\zeta)$ in (4.6), one can see that

$$P_{N_1} S(0)|_{N_2} = P_{N_1} A|_{N_2} = 0 .$$

From this relation the invertibility of the operator $(I_{E_2 \oplus N_1} - S|_{N_2} \omega P_{N_1})$ follows.

3) Using the block decomposition of $S(\zeta)$

$$S = \begin{bmatrix} s_0 & s_2 \\ s_1 & s \end{bmatrix} : \begin{bmatrix} E_1 \\ N_2 \end{bmatrix} \longrightarrow \begin{bmatrix} E_2 \\ N_1 \end{bmatrix}$$

one can rewrite formula (4.8) in the form

$$\omega = s_0 + s_2(I_{N_2} - \varepsilon s)^{-1} \varepsilon s_1 . \tag{4.10}$$

5. THE GENERALIZED BI-TANGENTIAL SCHUR-NEVANLINNA-PICK (SNP) PROBLEM

We use the following notations:

$H^{\infty}[E_1, E_2]$ is the set of holomorphic (on $\mathbb{D}$) o.f. $h(\zeta)$ whose values are bounded linear operators from E_1 to E_2, where E_1 and E_2 are separable Hilbert Spaces, and $\|h\|_\infty = \sup_{\zeta \in \mathbf{D}} \|h(\zeta)\| < \infty$. $\mathbf{B}[E_1, E_2]$ is the unit ball in $H^\infty[E_1, E_2]$. The following problem was studied by D.Z. Arov [6, 7]. Results concerning the so-called completely indeterminate case were exposed there. In this section we shall survey a number of results from [22, 23].

THE SNP PROBLEM. *Let the o.f. $\omega_0 \in \mathbf{B}[E_1, E_2]$, the $*$-inner function $\theta_1 \in \mathbf{B}[E_1, \widetilde{E}_1]$ (i.e., $\theta_1 \theta_1^* = I_{\widetilde{E}_1}$ a.e. on $\mathbf{T}$), and the inner function $\theta_2 \in \mathbf{B}[\widetilde{E}_2, E_2]$ (i.e., $\theta_2^* \theta_2 = I_{\widetilde{E}_2}$ a.e. on $\mathbf{T}$) be given.*

It is required to describe all the o.f. $\omega \in \mathbf{B}[E_1, E_2]$ such that

$$\omega - \omega_0 = \theta_2 h \theta_1 , \quad h \in H^\infty[\widetilde{E}_1, \widetilde{E}_2] . \tag{5.1}$$

EXAMPLE. Let θ_1 and θ_2 be scalar Blaschke products such that $\theta_1 \theta_2$ has simple zeros $\{\zeta_j\}$, then the condition (5.1) means that the values of ω and ω_0 coincide on ζ_j. Hence we have the classical Nevanlinna-Pick problem with data $\{\zeta_j\}$ and $\{\omega_j\}$, where $\omega_j = \omega_0(\zeta_j)$.

This example shows that knowledge of the whole function ω_0 is redundant for the statement of the SNP-Problem. It appears that also in the general case only "part" of the function ω_0 contains essential information for the SNP problem.

Let

$$K_{1,*} = L^2(E_1) \ominus \theta_1^* H_-^2(\widetilde{E}_1) , \quad K_2 = L^2(E_2) \ominus \theta_2 H_+^2(\widetilde{E}_2) , \tag{5.2}$$

where as usual L^2 denotes the space of square summable vector functions on $\mathbf{T}$ and H^2 denotes the Hardy space.

PROPOSITION 5.1. *Every o.f. $\omega \in H^\infty[E_1, E_2]$ defines a bounded operator $W : K_{1,*} \to K_2$ by the formula*

$$W = P_{K_2} \omega |_{K_{1,*}} .$$

If $\omega \in \mathbf{B}[E_1, E_2]$, then W is a contraction.

PROPOSITION 5.2. *Condition (5.1) holds iff*

$$P_{K_2} \omega |_{K_{1,*}} = P_{K_2} \omega_0 |_{K_{1,*}} \tag{5.3}$$

i.e., iff the operators defined by the functions ω and ω_0 coincide.

Next, we will reformulate the generalized bi-tangential problem and include it into the general scheme of the AIP. To this end we introduce some new objects.

Since $K_{1,*}$ is invariant under the operator of multiplication by the independent variable t, the operator $T_{K_{1,*}} = t|_{K_{1,*}}$ can be defined on $K_{1,*}$. Similarly the operator $T_{K_2} = \bar{t}|_{K_2}$ can be defined on K_2.

PROPOSITION 5.3. *For every function* $\omega \in H^\infty[E_1, E_2]$, *the operator* $W = P_{K_2}\omega|_{K_{1,*}}$ *possesses the following property:*

$$W T_{K_{1,*}} = T_{K_2} W .\tag{5.4}$$

Now let

$$K_{\theta_1^*} = H_-^2(E_1) \ominus \theta_1^* H_-^2(\tilde{E}_1) , \quad K_{\theta_2} = H_+^2(E_2) \ominus \theta_2 H_+^2(\tilde{E}_2) .$$

Then

$$K_{1,*} = K_{\theta_1^*} \oplus H_+^2(E_1) , \qquad K_2 = H_-^2(E_2) \oplus K_{\theta_2} .\tag{5.5}$$

According to this decomposition of spaces, we consider the decompositions

$$T_{K_1} = \begin{bmatrix} T_{\theta_1^*} & 0 \\ * & * \end{bmatrix} , \quad T_{K_2}^* = \begin{bmatrix} * & * \\ 0 & T_{\theta_2}^* \end{bmatrix} ,$$

where $T_{\theta_1^*} = P_- t|_{K_{\theta_1^*}}$, $T_{\theta_2}^* = P_+ \bar{t}|_{K_{\theta_2}}$.

PROPOSITION 5.4. *According to the decompositions (5.5), the operator* W *has the triangular form*

$$W = \begin{bmatrix} W_1 & 0 \\ W_{21} & W_2 \end{bmatrix} .\tag{5.6}$$

We are now able to reformulate the generalized bi-tangential problem.

THE SNP PROBLEM. *For a given operator* $W : K_{1,*} \to K_2$ *such that*

a) *W is a contraction,*

b) $W = \begin{bmatrix} W_1 & 0 \\ W_{21} & W_2 \end{bmatrix}$ *in the decomposition of the spaces (5.5), and*

c) $W T_{K_{1,*}} = T_{K_2} W,$

describe all the holomorphic (on $\mathbf{D}$*) contractive operator-valued functions* $\omega(\zeta) : E_1 \to E_2$ *such that*

$$P_{K_2}\omega|_{K_{1,*}} = W .$$

For the rest of this section we shall work in the following setting:

Let

$$X = \left\{ x = \begin{bmatrix} x_1 \\ x_2 \end{bmatrix} : x_1 \in K_{\theta_1^*}, \ x_2 \in K_{\theta_2} \right\},$$

and set

$$D = \begin{bmatrix} I - W_1^* W_1 & W_{21}^* \\ W_{21} & I - W_2 W_2^* \end{bmatrix}. \tag{5.7}$$

Then D defines a quadratic form on X: $\langle Dx, x \rangle$, where $\langle \ , \ \rangle$ is an inner product in $K_{\theta_1^*} \oplus K_{\theta_2}$.

PROPOSITION 5.5. *The operator W is contractive if and only if the form D is nonnegative.*

PROPOSITION 5.6. *The following identity holds:*

$$\langle DT_2 x, T_2 y \rangle - \langle DT_1 x, T_1 y \rangle = \langle M_1 x, M_1 y \rangle_{E_1} - \langle M_2 x, M_2 y \rangle_{E_2}, \tag{5.8}$$

where

$$T_1 = \begin{bmatrix} T_{\theta_1^*} & 0 \\ 0 & I \end{bmatrix}, \quad T_2 = \begin{bmatrix} I & 0 \\ 0 & T_{\theta_2}^* \end{bmatrix},$$

$$M_1 x = t x_1 \big|_0 + W_2^* x_2 \big|_0 : \ X \longrightarrow E_1,$$

$$M_2 x = t W_1 x_1 \big|_0 + x_2 \big|_0 : \ X \longrightarrow E_2. \tag{5.9}$$

The notation $\big|_0$ designates the zero Fourier coefficient of the L^2-vector function. This identity permits us to include the SNP Problem in the abstract scheme. We consider the following problem:

THE SNP' PROBLEM. *Let W be the same operator as in the SNP problem. Let D be defined by formula (5.7), $X = K_{\theta_1^*} \oplus K_{\theta_2}$, and let T_1, T_2, M_1, M_2 be defined by formulas (5.9). It is required to describe all the functions $\omega \in \mathbb{B}[E_1, E_2]$ such that there exists a map $F : X \to H_\omega$ with the following properties:*

For every $x \in X$

a) $FT_2 x = t \cdot FT_1 x - \begin{bmatrix} I_{E_2} & \omega \\ \omega^* & I_{E_1} \end{bmatrix} \begin{bmatrix} -M_2 x \\ M_1 x \end{bmatrix}$ *for a.e. $t \in \mathbb{T}$,* $\tag{5.10}$

b) $\|Fx\|_{H_\omega}^2 \le \langle Dx, x \rangle.$ $\tag{5.11}$

THEOREM 5.7.

1) The set of solutions of the SNP' problem and the SNP problem coincide.

2) For every solution ω, there is a unique map $F : X \rightarrow H_\omega$ which satisfies (5.10) and (5.11); it is defined by the formula

$$F^\omega x = \begin{bmatrix} I_{E_2} & \omega \\ \omega^* & I_{E_1} \end{bmatrix} \begin{bmatrix} x_2 - W_1 x_1 \\ -W_2^* x_2 + x_1 \end{bmatrix} \tag{5.12}$$

3) For every solution ω and every $x \in X$,

$$\|F^\omega x\|_{H_\omega}^2 = \langle Dx, x \rangle .$$

Since the SNP$'$ problem fits into the scheme of the AIP, all the constructions from Section 4 are applicable. The isometry V (see (4.1)) and the s.o.f. of the SNP$'$ problem

$$S = \begin{bmatrix} s_0 & s_2 \\ s_1 & s \end{bmatrix} \in \mathbb{B}[E_1 \oplus N_2, \ E_2 \oplus N_1] .$$

(see (4.6)) are defined from the data of the problem. It follows from the theorem of Section 4:

THEOREM 5.8. *The set of all solutions ω of the SNP problem can be described as follows:*

$$\omega = s_0 + s_2 \varepsilon (I_{N_1} - s\varepsilon)^{-1} s_1 \tag{5.13}$$

where ε is an arbitrary function in $\mathbb{B}[N_1, N_2]$.

We now list some specific properties of the s.o.f. of the SNP problem.

PROPOSITION 5.9. *Let S be a scattering operator for the SNP problem. Then*

$$a) \quad s_1 = \widetilde{s}_1 \theta_1 , \quad and \quad b) \quad s_2 = \theta_2 \widetilde{s}_2 , \tag{5.14}$$

where $\widetilde{s}_1$ is an outer o.f. and $\widetilde{s}_2$ is a $*$-outer o.f. (here we use the terminology of [40]).

REMARK. Properties (5.14) conform well with the setting of the SNP problem in the form (5.1). In fact, one can see from (5.13) and (5.14) that

$$\omega - s_0 = \theta_2 h \theta_1 ,$$

where $h = \widetilde{s}_2 \varepsilon (1_{N_1} - s\varepsilon)^{-1} \widetilde{s}_1 \in H^\infty[\widetilde{E}_1, \widetilde{E}_2]$.

We now discuss another property of the matrix S, which generalizes the following famous

THEOREM (R. NEVANLINNA). *If a scalar Nevanlinna-Pick interpolation problem (with an infinite set of interpolation nodes) has a nonunique solution, then it has a unimodular one [33].*

The Nevanlinna theorem permits an equivalent formulation which is more convenient for generalizations (see [1, 3,]): *If a scalar Nevanlinna-Pick interpolation problem (with an infinite set of interpolation nodes) has a nonunique solution, then the s.o.f. of the SNP problem S is a 2×2 matrix with unitary boundary values.*

THEOREM 5.10. *Let S be the s.o.f. of a SNP problem. Then*

$$\begin{bmatrix} I_{N_2} & s^* \\ s & I_{N_1} \end{bmatrix} = \begin{bmatrix} s_2^* & 0 \\ 0 & s_1 \end{bmatrix} \begin{bmatrix} I_{E_2} & s_0 \\ s_0^* & I_{E_1} \end{bmatrix}^{[-1]} \begin{bmatrix} s_2 & 0 \\ 0 & s_1^* \end{bmatrix} \tag{5.15}$$

for a.e. $t \in \mathbb{T}$.

The expression on the right hand side of (5.15) is equal to C^*C, where

$$C = \begin{bmatrix} I_{E_2} & s_0 \\ s_0^* & I_{E_1} \end{bmatrix}^{-\frac{1}{2}} \begin{bmatrix} s_2 & 0 \\ 0 & s_1^* \end{bmatrix}$$

(see Section 3).

This theorem follows from the equality $\|F^\omega x\|_{H_\omega}^2 = \langle Dx, x \rangle$ for every $x \in X$ and every solution ω [24].

THEOREM 5.11. *Let S be the s.o.f. of a SNP problem with* $\dim E_1 < \infty$ *and* $\dim E_2 < \infty$. *Then the condition (5.15) may be rewritten in each of the following three forms (for a.e.* $t \in \mathbb{T}$):

$$1) \quad \mathrm{rank}\begin{bmatrix} I & S \\ S^* & I \end{bmatrix} = \mathrm{rank}\begin{bmatrix} I & s_0 \\ s_0^* & I \end{bmatrix}$$

$$2) \quad \mathrm{rank}(I - S^*S) = \mathrm{rank}(I - s_0^*s_0) - \dim N_1 \tag{5.16}$$

$$3) \quad \mathrm{rank}(I - SS^*) = \mathrm{rank}(1 - s_0 s_0^*) - \dim N_2 .$$

Moreover, if $\varepsilon \in \mathbb{B}[N_1, N_2]$ *and* ω *is the solution of the SNP Problem corresponding to* ε *by formula (5.13), then the following equivalent equalities hold (for a.e.* $t \in \mathbb{T}$):

$$1) \quad \mathrm{rank}\begin{bmatrix} I & S \\ S^* & I \end{bmatrix} = \mathrm{rank}\begin{bmatrix} I & \omega \\ \omega^* & I \end{bmatrix} + \dim \ker \begin{bmatrix} I & \varepsilon \\ \varepsilon^* & I \end{bmatrix} ;$$

$$2) \quad \mathrm{rank}(I - S^*S) = \mathrm{rank}(I - \omega^*\omega) - \mathrm{rank}(I - \varepsilon^*\varepsilon) ; \tag{5.17}$$

$$3) \quad \mathrm{rank}(I - SS^*) = \mathrm{rank}(I - \omega\omega^*) - \mathrm{rank}(I - \varepsilon\varepsilon^*) ;$$

(where 'rank' means the dimension of the image and 'dim ker' means the dimension of the kernel). If $\varepsilon = 0$, *then the equalities (5.17) reduce to (5.16).*

The following theorem is a corollary of the previous one:

THEOREM 5.12. *Let* dim $E_1 < \infty$, dim $E_2 < \infty$ *and let S be a scattering operator of the SNP problem. Then:*

1) dim $N_1 \leq$ dim $\tilde{E}_1$, dim $N_2 \leq$ dim $\tilde{E}_2$.

2) *If* dim $N_1 =$ dim E_1 *(i.e. the right rank of indefiniteness is maximal), then*

$$I - S^*S = 0 \quad a.e. \text{ on } \mathbf{T}.$$

3) *If* dim $N_2 =$ dim E_2 *(i.e. the left rank of indefiniteness is maximal), then*

$$I - SS^* = 0 \quad a.e. \text{ on } \mathbf{T}.$$

4) *If* dim $N_2 \geq$ dim $N_1 =$ dim E_1, *then among the solutions of the SNP problem there are inner ones, i.e., ω such that*

$$I - \omega^*\omega = 0 \quad a.e. \text{ on } \mathbf{T}.$$

5) *If* dim $N_1 \geq$ dim $N_2 =$ dim E_2, *then among the solutions of the SNP problem there are *-inner ones, i.e., ω such that*

$$I - \omega\omega^* = 0 \quad a.e. \text{ on } \mathbf{T}.$$

REMARKS.

1) Assertions 4 and 5 of the last theorem yield the theorem of R. Nevanlinna which was mentioned above, when dim $E_1 =$ dim $E_2 = 1$, and θ_1, θ_2 are scalar Blaschke products.

2) All the statements of Theorem 5.12 were proved in the works of Adamjan, Arov, Krein [1, 3] and Adamjan [2]. Thus Theorems 5.10 and 5.11 may be regarded as a generalization of these results to the semi-determinate case.

3) The complete characterization of the s.o.f. of the SNP problem was obtained in [6, 7] for the completely indeterminate case and in [22] for the general case.

6. INNER-OUTER FACTORIZATION OF J-CONTRACTIVE MATRIX-FUNCTIONS

6.1. Definitions and notation.

a) This section deals only with matrix-functions (m.f.), which means that all the coefficient spaces E, N, etc., are Euclidean.

b) Let a Euclidean space K be decomposed into an orthogonal sum of subspaces $K = K_1 \oplus K_2$. With this decomposition we associate the operator

$$J_K = J_{K_1 \oplus K_2} = P_{K_1} - P_{K_2} . \qquad (6.1)$$

c) The class $\mathbb{B}_J[E_1 \oplus E_2, N_1 \oplus N_2]$ consists of meromorphic m.f. $\mathcal{U}(\zeta)$ on $\mathbb{D}$ such that $\mathcal{U}(\zeta) \in [E, N]$ for each regular point ζ (where $E = E_1 \oplus E_2$, $N = N_1 \oplus N_2$) which satisfy the following properties:

$$J_E - \mathcal{U}^*(\zeta) J_N \mathcal{U}(\zeta) \geq 0 , \qquad J_N - \mathcal{U}(\zeta) J_E \mathcal{U}^*(\zeta) \geq 0 . \qquad (6.2)$$

The left-hand sides in (6.2) are called the J-forms of the matrix $\mathcal{U}(\zeta)$.

REMARK. The m.f. $\mathcal{U}(\zeta) \in \mathbb{B}_J[E_1 \oplus E_2, N_1 \oplus N_2]$ may be represented as the linear-fractional transformation of a m.f. $S(\zeta) \in \mathbb{B}[E_1 \oplus N_2, E_2 \oplus N_1]$

$$\mathcal{U}(\zeta) = (P_{N_1} S(\zeta) + P_{N_2})(P_{E_1} + P_{E_2} S(\zeta))^{-1} . \qquad (6.3)$$

This transform is called the Potapov-Ginzburg transform. Note that the existence of boundary values of $\mathcal{U}(t)$ for a.e. $t \in \mathbb{T}$ follows from (6.3).

d) The m.f. $\mathcal{U}(\zeta) \in \mathbb{B}_J[E_1 \oplus E_2, N_1 \oplus N_2]$ is called J-inner if the equality

$$J_E - \mathcal{U}^*(t) J_N \mathcal{U}(t) = 0 , \qquad t \in \mathbb{T} ,$$

takes place at a.e. point t of the boundary.

e) Our definition of a J-outer m.f. is close to the characteristic property of outer contractive operator-functions [40: V, Sect. 4].

A m.f. $\mathcal{U}(\zeta) \in \mathbb{B}_J[E_1 \oplus E_2, N_1 \oplus N_2]$ is called J-outer if it has the following properties:

(e1) Every m.f. $\mathcal{B}(\zeta) \in \mathbb{B}_J[E_1 \oplus E_2, K_1 \oplus K_2]$ satisfying the inequality

$$J_E - \mathcal{B}^*(t) J_K \mathcal{B}(t) \geq J_E - \mathcal{U}^*(t) J_N \mathcal{U}(t) , \qquad t \in \mathbb{T} , \qquad (6.4)$$

can be represented in the form

$$B(\zeta) = C(\zeta)\mathcal{U}(\zeta) \tag{6.5}$$

where $C(\zeta) \in \mathbf{B}_J[N_1 \oplus N_2, K_1 \oplus K_2]$.

(e2) There exists a point $\zeta_0 \in \mathbb{D}$ such that the image of $\mathcal{U}(\zeta_0)$ coincides with the whole of N, i.e., Ran $\mathcal{U}(\zeta_0) = N_1 \oplus N_2$.

REMARKS.

1. There are other definitions of a J-outer m.f. [14, 32, 36]. Note in particular that the spectrum of a J-outer m.f. $\mathcal{U}(\zeta)$ as defined here, is absolutely continuous [32].

2. If $\mathcal{U}(\zeta) \in \mathbf{B}_J[E_1 \oplus E_2, N_1 \oplus N_2]$ is J-outer and J-inner simultaneously, then $\mathcal{U}(\zeta)$ is a J-unitary constant matrix.

3. If a m.f. $\mathcal{U}(\zeta)$ satisfies only condition (e1), then it is of the form $\mathcal{U}(\zeta) = C_0\mathcal{U}_0(\zeta)$, where $\mathcal{U}_0(\zeta)$ is a J-outer m.f. and C_0 is a J-isometric constant.

6.2. Theorem on J-inner-outer factorization.

THEOREM 6.1. *Let $B(\zeta) \in \mathbf{B}_J[E_1 \oplus E_2, K_1 \oplus K_2]$. Then it can be represented in the form*

$$B(\zeta) = C(\zeta)\mathcal{U}(\zeta) , \tag{6.6}$$

where $\mathcal{U}(\zeta) \in \mathbf{B}_J[E_1 \oplus E_2, N_1 \oplus N_2]$ is a J-outer m.f., and
$C(\zeta) \in \mathbf{B}_J[N_1 \oplus N_2, K_1 \oplus K_2]$ is a J-inner m.f.

Our proof of Theorem 6.1 is based on splitting-off the J-outer factor. We outline here the steps of this proof.

Let us define a nonnegative measurable m.f. $\Gamma(t)$, $t \in \mathbf{T}$, by the equality

$$\Gamma(t) = J_E - B^*(t)J_K B(t) , \qquad t \in \mathbf{T} \tag{6.7}$$

and consider the following:

PROBLEM 6.1. Describe every m.f. $B'(\zeta) \in \mathbf{B}_J[E_1 \oplus E_2, K_1' \oplus K_2']$, which satisfies the inequality

$$J_E - B'^*(t)J_{K'}B'(t) \geq \Gamma(t) , \qquad t \in \mathbf{T} . \tag{6.8}$$

The main point of our construction is the fact that the *solution of Problem 6.1 can be obtained by means of the AIP scheme*. In subsection 6.3 we will touch upon this subject examining a more general Problem 6.2. Now we use the Main Theorem of Section 4 on the description of the solutions of the AIP. In our case it yields the following:

THEOREM 6.2. *The general solution of Problem 6.1 has the form*

$$B'(\zeta) = C'(\zeta)\mathcal{U}(\zeta) , \quad C'(\zeta) \in \mathbb{B}_J[N_1 \oplus N_2, K_1' \oplus K_2'] . \tag{6.9}$$

Moreover, the m.f. $\mathcal{U}(\zeta)$ *is the Potapov-Ginzburg transform of the s.o.f.* $S(\zeta)$ *of some isometry* V *that can be constructed entirely from the given m.f.* $\Gamma(t)$. *The intermediate spaces* N_1 *and* N_2 *are the defect subspaces of* V. *The boundary values of the* J-*form of the m.f.* $\mathcal{U}(\zeta)$ *coincide with* $\Gamma(t)$:

$$J_E - \mathcal{U}^*(t)J_N\mathcal{U}(t) = \Gamma(t) , \quad t \in \mathbf{T} .$$

We would like to indicate that the assertion of Theorem 6.2 implies that $\mathcal{U}(\zeta)$ satisfies the main property (e1) from the definition of a J-outer m.f. The property (e2) is also fulfilled (mainly because the subspaces N_1 and N_2 are defect spaces). Thus the matrix $\mathcal{U}(\zeta)$ from Theorem 6.2 is J-outer.

We would like to recall that the original m.f. $B(\zeta)$ is a solution of Problem 6.1 since its J-form determined the m.f. $\Gamma(t)$. Therefore $B(\zeta)$ may be represented in the form (6.9), i.e.,

$$B(\zeta) = C(\zeta)\mathcal{U}(\zeta) , \quad C(\zeta) \in \mathbb{B}_J[N_1 \oplus N_2, K_1 \oplus K_2] . \tag{6.10}$$

The coincidence of the boundary J-forms of the J-outer m.f. $\mathcal{U}(\zeta)$ and the m.f. $B(\zeta)$ implies that the factor $C(\zeta)$ is J-inner. Hence (6.10) is the desired J-inner-outer factorization of $B(\zeta)$.

6.3. Problems on disks.

We would like to extend the statement of Problem 6.1.

Firstly, making use of the Potapov-Ginzburg transform we pass from J-contractive m.f.'s to contractive ones. Let

$$B'(\zeta) = (P_{K_1'}R(\zeta) + P_{K_2'})(P_{E_1} + P_{E_2}R(\zeta))^{-1} , \tag{6.11}$$

where $R(\zeta) \in \mathbb{B}[E_1 \oplus K_2', E_2 \oplus K_1']$. Then inequality (6.8) takes the form

$$I - R^*(t)R(t) \geq (P_{E_1} + P_{E_2}R(t))^*\Gamma(t)(P_{E_1} + P_{E_2}R(t)) . \tag{6.12}$$

Next, we pass from the nonnegative m.f. $\Gamma(t)$ to the bounded matrix-function $\Lambda(t) = \{\Gamma(t)/(I + \Gamma(t))\}^{\frac{1}{2}}$. Then $\Gamma(t) = \Lambda(t)(I - \Lambda^2(t))^{-1}\Lambda(t)$ and the inequality (6.12) can be rewritten in the following way:

$$I - R^*(t)R(t) \geq (P_{E_1} + P_{E_2}R(t))^*\Lambda(t)(I - \Lambda^2(t))^{-1}\Lambda(t)(P_{E_1} + P_{E_2}R(t)) . \tag{6.13}$$

But this in turn is equivalent to the nonnegativity of a block-matrix:

$$
\begin{bmatrix}
I - \Lambda^2(t) & \vdots & \Lambda(t)P_{E_1} + \Lambda(t)P_{E_2}R(t) \\
\text{--} & \vdots & \text{--} \\
* & \vdots & I - R^*(t)R(t)
\end{bmatrix}
\geq 0 . \qquad (6.14)
$$

Removing the requirement of the invertibility of the matrix $I - \Lambda^2(t)$ we come to the following problem.

PROBLEM 6.2 (ON DISKS). *Let $\Lambda(t)$ be a measurable m.f., $\Lambda(t) \in [E, E]$ for a.e. $t \in \mathbb{T}$, $0 \leq \Lambda(t) \leq 1$. It is required to describe the set of all holomorphic contractive $[E_1 \oplus K_2', E_2 \oplus K_1']$-valued m.f. $R(\zeta)$ which satisfy the inequality (6.14).*

Let us explain our terminology. For a.e. $t \in \mathbb{T}$, the boundary values of the m.f. $R(t)$ belong to the matrix disks given in the statement of Problem 6.2. A similar statement is presented in [14].

The applicability of the general method of solving of interpolation problems which was presented above to Problem 6.2 is based on the following two assertions [42, 43].

THEOREM 6.3. *Problem 6.2 is solvable iff the quadratic form D:*

$$
D(x, x) = \langle (1 - \Lambda^2 - \Lambda P_+ J_E \Lambda)x, x \rangle \qquad (6.15)
$$

is nonnegative. Here $x \in L^2(E)$, P_+ is the orthoprojection on $H_+^2(E)$, and $\langle \, , \, \rangle$ is the inner product in $L^2(E)$.

PROPOSITION. *Let T be the operator of multiplication by the independent variable in $L^2(E)$, $(Tx)(t) = tx(t)$, and let $D(x, x)$ be the quadratic form defined by (6.15). Then the following identity holds:*

$$
D(x, x) - D(Tx, Tx) = \langle M_1 x, M_1 x \rangle_{E_1} - \langle M_2 x, M_2 x \rangle_{E_2} , \qquad (6.16)
$$

where

$$
M_1 x = P_{E_1} P_+ T P_- \Lambda x , \qquad M_2 x = P_{E_2} P_+ T P_- \Lambda x
$$

are the operators from $L^2(E)$ to E_1 and E_2, respectively.

REFERENCES

[1] V.M. Adamjan, D.Z. Arov and M.G. Krein: *Infinite block Hankel matrices and related continuation problems.* Amer. Math. Soc. Transl. 111:2(1978).

[2] V.M. Adamjan: *Non-degenerate unitary intertwinings of semi-unitary operators* (Russian), Funkt. Anal. i Prilozhen, 7:4 (1973), 1-16; English transl. Functional Anal. Appl. 7 (1973), 255-267.

[3] V.M. Adamjan, D.Z. Arov and M.G. Krein: *Infinite Hankel matrices and generalized Caratheodory-Fejer and I. Schur problems.* (Russian), Funkt. Anal. i Prilozhen, 2:4(1968), 1-17.

[4] V.M. Adamjan, D.Z. Arov and M.G. Krein: *Bounded operators that commute with a contraction of class C_{00} of unit rank of non-unitary.* (Russian), Funkt. Anal. i Prilozhen, 3:3(1969), 86-87.

[5] N.I. Akhiezer: *The Classical Moment Problem.* Oliver and Boyd, Edinburgh, 1965.

[6] D.Z. Arov: *Regular J-inner matrix-functions and related continuation problems.* Operator Theory: Advances and Applications, 43 (1990), 63-87.

[7] D.Z. Arov: *γ-derived matrices, J-inner matrix-functions and related extrapolation problems.* (Russian), Teor. Funk., Funkt. Anal. i ikh Prilozhen, (Kharkov), 51 (1989), 61-67; 52 (1989), 103-109; 53 (1990), 57-65; English transl. J. Soviet Math. 52 (1990), 3487-3491; 52 (1990) 3421-3425; 58 (1992), 532-537.

[8] D.Z. Arov and L.Z. Grossman: *Scattering matrices in the extension theory of isometrical operators.* Soviet Math. Dokl., 27 (1983), 573-578.

[9] J.A. Ball and J.W. Helton: *A Beurling-Lax theorem for Lie group $U(m,n)$ which contains most classical interpolation problems.* J. Oper. Theory 9 (1983), 107-142.

[10] L. de Branges: *Hilbert Spaces of Entire Functions.* Prentice Hall, Englewood Cliffs, NJ, 1968.

[11] L. de Branges and J. Rovnyak: *Canonical models in quantum scattering theory.* In: "Perturbation Theory and Its Application in Quantum Mechanics", Wiley, New York, 359-391, 1966.

[12] H. Dym: *J contractive matrix functions, reproducing kernel Hilbert spaces and interpolation.* CBMS Regional Conf. Ser. in Math., 71, Amer. Math. Soc., Providence, R.I., 1989.

[13] A.V. Efimov and V.P. Potapov: *J-expansive matrix-functions and their role in the analytic theory of electrical circuits.* (Russian), Uspehi Mat. Nauk 28 (1974), 65-130; English transl. in Russian Math. Surveys, 28 (1974).

[14] J.W. Helton: *Operator theory, analytic functions, matrices and electrical engineering*. CBMS Regional Conf., Series in Math., No. 68, American Math. Soc., Providence, R.I., 1987.

[15] T.S. Ivanchenko and L.A. Sahnovich: *Operator approach to investigation of interpolation problems*. (Russian), 1985, deposited in Ukr NIINTI, No. 701-Uk D85.

[16] T.S. Ivanchenko and L.A. Sahnovich: *Operator-theoretic approach to the V. P. Potapov scheme of investigation of interpolation problems*. (Russian), Ukr. Math. Jour., 39:5 (1987), 573-578.

[17] V.E. Katsnelson: *Continuous analogs of Hamburger-Nevanlinna theorem and fundamental matrix inequality of classical problems*. (Russian), Teor. Funk., Funkt. Anal. i ikh Prilozhen. (Kharkov), 36 (1981), 31-48; 37 (1982), 31-58; 39 (1983), 61-73; 40 (1983), 79-80; English transl. in Amer. Math. Soc. Transl. (2) 136 (1987), 49-108.

[18] V.E. Katsnelson: *The fundamental matrix inequality of the problem of positive-definite kernel decomposition on the elementary products*. (Russian), Dokl. Akad. Nauk Ukr. SSSR, Ser. A 2 (1984), 10-13.

[19] V.E. Katsnelson: *Regularization of the fundamental matrix inequality of the problem of positive-definite kernel decomposition on the elementary products*. (Russian), Dokl. Akad. Nauk Ukr. SSSR, Ser. A, 3 (1984), 6-8.

[20] V.E. Katsnelson, A. Ya. Kheifets and P.M. Yuditskii: *The abstract interpolation problem and extension theory of isometric operators*. (Russian), in: Operator in Spaces of Functions and Problems in Function Theory: Collected scientific papers, Kiev, Naukova Dumka, 1987, 83-96.

[21] A.Ya. Kheifets: *Parseval equality in abstract interpolation problem and coupling of open systems*. (Russian), Teor. Funk., Funkt. Anal. i ikh Prilozhen, (Kharkov), 49 (1988), 112-120; 50 (1988), 98-103; English transl. in Journal of Soviet Math., 49 (1990), 1114-1120; N6, 1307-1310.

[22] A.Ya. Kheifets: *Generalized bi-tangential Schur-Nevanlinna-Pick problem, related Parseval equality and scattering operator*. (Russian), deposited in VINITI 11.05.1989, No. 3108-889 Dep., 1-60, 1989.

[23] A.Ya. Kheifets: *Generalized bi-tangent Shur-Nevanlinna-Pick problem and related Parseval equality*. (Russian), Teor. Funk., Funkt. Anal. i ikh Prilozhen, (Kharkov), 54 (1990), 89-96; English transl. in Journal of Soviet Math.

[24] A.Ya. Kheifets: *Nevanlinna-Adamjan-Arov-Krein theorem in semi-determined case*. (Russian), Teor. Funk., Funkt. Anal. i ikh Prilozhen (Kharkov), 56 (1991), 128-137; English transl. in Journal of Soviet Math.

[25] A. Ya. Kheifets and P.M. Yuditskii: *Interpolation of operator commuting with shorted shift by I. Schur class functions.* (Russian), Teor. Funk., Funkt. Anal. i ikh Prilozhen (Kharkov), 40 (1983), 129-136.

[26] I.V. Kovalishina: *Analytic theory of a class of interpolation problems.* (Russian), Math. SSSR Izv., 22 (1984), 419-463.

[27] I.V. Kovalishina and V.P. Potapov: *Indefinite metric in Nevanlinna-Pick problem.* (Russian), Dokl. Akad. Nauk Armjan. - SSR, ser. mat., 9:1(1974), 3-9; Private translation by T. Ando, Sapporo, 1982.

[28] M.G. Krein: *On generalized moment problem.* (Russian), Dokl. Akad. Nauk USSR, 44:6(1944), 233-243.

[29] M.G. Krein: *On Hermitian operators with direction functionals.* (Russian), Sborn. Trudov Inst. Mat. Ukr., 10 (1948), 83-105.

[30] M.G. Krein: *Basic considerations on the representation theory of Hermitian operators with defect index (m, m).* (Russian), Ukr. Mat. Zhurnal, 1:2(1949), 3-66.

[31] M.S. Livshits: *On one application Hermitian operators theory to generalized moment problem.* (Russian), Dokl. Akad. Nauk. USSR, 44 (1944), 3-7.

[32] S.H. Naboko: *Absolutely continuous spectrum of non dissipative operator and functional model.* (Russian), Zap. Nauchn. Sem. LOMI, 65 (1976), 90-102; 73 (1977), 118-135.

[33] N.K. Nikolskii: *Treatise on the Shift Operator,* (with an Appendix by S.V. Hruscev and V.V. Peller), Springer Verlag, Heidelberg, 1986.

[34] N.K. Nikolskii and V.I. Vasyunin: *A unified approach to function models and the transcription problem.* Operator Theory: Advances and Application 41 (1989), 405-434.

[35] A.A. Nudelman: *On a generalization of classical interpolation problems.* Sov. Math. Dokl. 23 (1961), 125-128.

[36] V.P. Potapov: *Multiplicative structure of J-nonexpending matrix-function.* (Russian), Dokl. Mosk. Mat. Obsc. (1955), 125-236.

[37] M. Rosenblum and J. Rovnyak: *Hardy Classes and Operator Theory.* Oxford University Press, 1986.

[38] D. Sarason: *Generalized interpolation in H^∞.* Trans. Amer. Math. Soc. 127:2 (1967), 179-203.

[39] D. Sarason: *Operator theoretic aspect of the Nevanlinna-Pick interpolation problem in operator and function theory.* Lancaster Conference Notes, Reidel, Dordrecht, Higtham, Mass., 1985, 279-314.

[40]   B.Sz.-Nagy and C. Foias: *Harmonic analysis of operators in Hilbert space.* North-Holland, Amsterdam, 1970.

[41]   P.M. Yuditskii: *Lifting problem.* (Russian), deposited in Ukr. NIINTI 18.04.1983, no. 311-Uk-D83.

[42]   P.M. Yuditskii: *On one problem connected with Nehari problem.* (Russian), deposited in Ukr. NIINTI 25.11.1983, no. 1336-Uk-D83.

[43]   P.M. Yuditskii: *On reconstruction of j-contractive analytic matrix-function by its j-form of boundary values and related interpolation problem.* (Russian), Teor. Funk. Funkt. Anal. i ikh Prilozhen (Kharkov), 44 (1985), 141-143; English transl. in Journal of Soviet Math.

[44]   P.M. Yuditskii: *Inner-outer factorization of j-expanding invertible matrix-functions.* (Russian), Teor. Funk. Funkt. Anal. i ikh Prilozhen (Kharkov), 46 (1986), 132-136; English transl. in Journal of Soviet Math.

A. Ya. Kheifets
Mathematical Department
Kharkov State University
Kharkov 310077
Ukraine

P. M. Yuditskii
Mathematical Division of the Institut for Low
  Temperature Physics and Engineering
Lenin Av. 47, Kharkov 310164
Ukraine

Operator Theory:
Advances and Applications, Vol. 72
© 1994 Birkhäuser Verlag Basel

ON THE THEORY OF INVERSE PROBLEMS FOR THE CANONICAL DIFFERENTIAL EQUATION

M. G. Krein and I. E. Ovcharenko

For any integer n let us denote by $\mathcal{B}_n$ the set of square entire matrix functions $W(z)$ of order $2n$ of the complex variable z which have the following properties:

$$\frac{[W(z)^* J_n W(z) - J_n]}{z - \bar{z}} \leq 0, \quad \operatorname{Im} z \neq 0, \tag{1}$$

$$W(z)^* J_n W(z) - J_n = 0, \quad \operatorname{Im} z = 0, \tag{2}$$

where $J_n = \|J_{jk}\|_{j,k=1}^2$ is a square block matrix with the blocks $J_{11} = J_{22} = 0$, $J_{12} = -J_{21} = I_n$. Let us call a meromorphic matrix function *real* if its values on the real axis are matrices whose elements are real numbers. Let us denote the set of real matrix function of $\mathcal{B}_n$ by $\mathcal{B}_{n;r}$. A matrix function $W(z)$ of order $2n$ can be represented as the block 2×2 matrix $\|w_{jk}(z)\|_{j,k=1}^2$. The mappings

$$W(z) \rightarrow \begin{bmatrix} \cos\theta_1 I_n & \sin\theta_1 I_n \\ -\sin\theta_1 I_n & \cos\theta_1 I_n \end{bmatrix} \begin{bmatrix} w_{11}(z) & w_{12}(z) \\ w_{21}(z) & w_{22}(z) \end{bmatrix} \begin{bmatrix} \cos\theta_2 I_n & \sin\theta_2 I_n \\ -\sin\theta_2 I_n & \cos\theta_2 I_n \end{bmatrix} \tag{3}$$

(θ_1, θ_2 can be any real numbers) leave the classes $\mathcal{B}_n$ and $\mathcal{B}_{n;r}$ invariant.

In this paper we try to answer the following questions:

1) Let the entire matrix function $F(z)$ of order n be given. What is the criterion for the existence of the matrix function $W(z) \in \mathcal{B}_n$ which has F as its block?

2) Let the entire matrix functions $F(z)$ and $G(z)$ or order n be given. What is the criterion for the existence of the matrix function $W(z) \in \mathcal{B}_n$ which has $[F\ G]$ as its block row?

3) How can one describe all $W(z) \in \mathcal{B}_n$ having a given block or a given block row?

For matrix functions in the class $\mathcal{B}_{1;r}$ these problems were first posed and solved in [1], in connection with the study of what is called the indeterminate case of the Sturm-Liouville boundary-value problem. In the present paper, the results obtained in [1] for the class $\mathcal{B}_1$ will be generalized for the case of matrix functions from $\mathcal{B}_n$. Whatever the value of n, questions 1)–3) are directly connected with inverse problems for the canonical differential equation. The reason is that, according to the fundamental theorem of V. P. Potapov, [2], (see also [3]) the matrix function $W(z)$, normalized by the condition $W(0) = I_{2n}$, belongs to the class $\mathcal{B}_n$ exactly when it is a matrix of the monodromy of the canonical differential equation of phase dimension $2n$ with the signature matrices J_n. The present paper has some points in common with the theory of entire Hermitian operators with deficiency index (n, n) [4, 5], the theory of Hilbert spaces of entire functions [6], and [7].

Let us say that the $m \times m$ matrix function $F(z)$, holomorphic in $\mathbb{C}_+$ belongs to the class $\mathcal{R}_m$, if $\operatorname{Im} F(z) \geq 0$ for all $z \in \mathbb{C}_+$.

As usual let us put for a meromorphic matrix function, $F^*(z) = [F(\bar{z})]^*$. We will say that a meromorphic matrix function $\Phi(z)$, whose values are matrices of order $m \times m$, is in the class $\widehat{C}e_m$ if

$$\frac{\operatorname{Im} \Phi(z)}{\operatorname{Im} z} \geq 0, \quad \operatorname{Im} z \neq 0 .$$

We will use a theorem which directly generalizes the well-known theorem of N. G. Chebotarev for the matrix case.

Theorem 1 *For the meromorphic matrix function $\Phi(z)$ to belong to $\widehat{C}e_m$, it is necessary and sufficient for it to admit the absolutely convergent expansion*

$$\Phi(z) = c_0 + c_1 z - \frac{c_{-1}}{z} - z \sum_j \frac{M_j}{\alpha_j(z - \alpha_j)} ,$$

(4)

$$\sum_j \alpha_j^{-1} M_j < \infty ,$$

in which α_j are real numbers, and $M_j \geq 0$, $c_{-1} \geq 0$, $c_1 \geq 0$, $c_0 = c_0^*$ are matrices of order m $(j = 1, 2, \ldots)$.

Let us note that a matrix function $\Phi(z) \in \widehat{C}e_m$ is holomorphic in $\mathbb{C}_+ \cup \mathbb{C}_-$ and Hermitian on the real line. Conversely, if a meromorphic matrix function $\Phi(z)$ is such that

1) $\Phi|\mathbb{C}_+ \in \mathcal{R}_m$, and

2) $\Phi(x)$ is Hermitian $(-\infty < x < \infty)$,

then it belongs to the class $\widehat{C}e_m$, and consequently can be represented in the form (4).

Let us denote by $\mathcal{N}_n$ the class of all entire matrix functions $F(z)$ of order n, $\det F \not\equiv 0$ for which the following absolutely convergent expansion holds:

$$F^{-1}(z) = \frac{c_{-1}}{z} + c_0 + z \sum_j \frac{r_j}{\alpha_j(z - \alpha_j)} ,$$

(5)

$$\sum_j \alpha_j^{-2} \|r_j\| < \infty ,$$

where α_j are real numbers ($\neq 0$), and c_{-1}, c_0, r_j are matrices of order n ($j = 1, 2, \ldots$).

It is easy to generalize for the matrix case Theorem 4 from [8]; the result is:

Theorem 2 *Any matrix function $F \in \mathcal{R}_n$ has the following property*

$$\overline{\lim} \ \frac{\log \|F(z)\|}{|z|} \ < \ \infty \ ,$$

and

$$\int_{-\infty}^{\infty} \frac{\log^+ \|F(x)\|}{1 + x^2} dx \ < \ \infty \ . \tag{6}$$

We will need a special transformation of block matrices, which is close to that studied in [9]. If $A = \|a_{j,k}\|_1^2$ is a block matrix of order $2n$ with blocks of order n and $\det a_{21} \neq 0$, let us denote by $\Omega(A)$ the matrix

$$\Omega(A) = \begin{bmatrix} a_{11} a_{21}^{-1} & a_{11} a_{21}^{-1} a_{22} - a_{12} \\ a_{21}^{-1} & a_{21}^{-1} a_{22} \end{bmatrix} . \tag{7}$$

It is an important property of the transformation Ω that $\Omega(\Omega(A)) = A$, and if $W = \Omega(A)$ then

$$W^* J_n W - J_n \leq 0 \quad \Leftrightarrow \quad \frac{A - A^*}{i} \geq 0$$

$$W^* J_n W - J_n = 0 \quad \Leftrightarrow \quad A - A^* = 0 \ . \tag{8}$$

Theorem 3 *Let F be an entire matrix function of order n, with $\det F \not\equiv 0$. In order for it to be a block (with any j, k; $j, k = 1, 2$ given in advance), of some matrix function of the class $\mathcal{B}_{n;r}$ it is necessary and sufficient for F to belong to $\mathcal{R}_n$ and to be real.*

Let us prove the necessary part of the theorem. It is obvious from (3) that we can confine ourselves to considering the case $F(z) = W_{21}(z)$ ($W \in \mathcal{B}_n$).

From the equality $A(z) = \Omega(W(z))$ with $W \in \mathcal{B}_n$, $\det w_{21}(z) \not\equiv 0$ it follows that $A(z) \in \hat{\mathcal{C}}_{e_{2n}}$. According to Theorem 1, the matrix function $A(z)$ admits an absolutely convergent expansion (4) in which C and H are matrices of order $2n$. We get the desired expansion $F(z)$ if we consider the block $a_{21}(z)$ in the expansion (4) of the matrix function $A(z)$.

We can easily develop the sufficient part of Theorem 3 from the proof of Theorem 4.

Theorem 4 *Let F, G be entire matrix functions of order n, with $\det F \not\equiv 0$ and $\det G \not\equiv 0$. The matrices F and G form the block row $F = w_{21}$, $G = w_{22}$ of some matrix function $W \in \mathcal{B}_n$ if and only if the following conditions hold:*

1) $F^{-1}G \in \hat{C}_{em}$,

2) $F \in \mathcal{R}_n$,

3) $\det[G(x) + iF(x)] \neq 0, -\infty < x < \infty$,

4) $\sum_j \alpha_j^{-2}[E^{-1}(\alpha_j)]^* m_j E^{-1}(\alpha_j) < \infty$,

where α_j and m_j are taken from the expansion (4) of the matrix function $\Phi(z) = F^{-1}(z)G(z)$, and $E(z) = G(z) + iF(z)$.

If conditions 1)–4) are fulfilled, and in addition F and G are real, we can find a matrix $W \in \mathcal{B}_{n;r}$ which has $[F, G]$ as its block row.

Proof. The necessity of condition 1) and 2) is made obvious by the properties of the transformation Ω and Theorem 3. Note that the matrix function

$$D(z) = [w_{11}(z)i + w_{12}(z)][w_{21}(z)i + w_{22}(z)]^{-1} \in \mathcal{R}_n$$

and consequently admits the known integral representation

$$D(z) = \alpha + \beta z + \int_{-\infty}^{\infty} \left(\frac{1}{\lambda - z} - \frac{\lambda}{1 + \lambda^2} \right) d\sum(\lambda) \tag{9}$$

where $\sum(\lambda)$ is a Hermitian matrix of distributions for which

$$\int_{-\infty}^{\infty} \frac{d\sum(\lambda)}{1 + \lambda^2} < \infty. \tag{10}$$

Since $\left. \text{Im } D(z) \right|_{z=\lambda} = [E(\lambda)E^*(\lambda)]^{-1}$ in any real interval of regularity of the matrix function $D(z)$, then according to the Stieltjes inversion formula

$$\frac{d\sum(\lambda)}{d\lambda} = [E(\lambda)E^*(\lambda)]^{-1}, \tag{11}$$

where $E(z) = G(z) + iF(z) (z = \lambda + iy)$. In view of (10) and (11), $\det[G(x)+iF(x)] \neq 0, -\infty < x < \infty$.

Let us generate the matrix function $A(z) = \Omega(W(z)) \in \hat{C}_{e2n}$. The following absolutely convergent expansions hold for $\|A(z)\| = \|a_{jk}(z)\|_{j,k=1}^2$.

$$a_{11}(z) = \hat{c}_0 + \hat{c}_1 z - \frac{c_{-1}}{z} - z \sum_j \frac{t_j}{\alpha_j(z - \alpha_j)},$$

$$a_{12}(z) = \frac{s_{-1}^*}{z} + z \sum_j \frac{r_j^*}{\alpha_j(z - \alpha_j)},$$

$$a_{21}(z) = \frac{s_{-1}}{z} + z \sum_j \frac{r_j}{\alpha_j(z - \alpha_j)}, \tag{12}$$

$$a_{22}(z) = c_0 + c_1 z - \frac{c_{-1}}{z} - z \sum_j \frac{m_j}{\alpha_j(z - \alpha_j)}.$$

From the equalities $F^{-1}(z)F(z) = F(z)F^{-1}(z) = I_n$ and $a_{22}(z) = F^{-1}(z)G(z)$ follows that $m_j = r_j G(\alpha_j)$; $c_{-1} = s_{-1}G(0)$; $r_j F(\alpha_j) = F(\alpha_j)r_j = 0$; $s_{-1}F(\alpha_j) = 0$, hence

$$r_j = m_j E^{-1}(\alpha_j), \quad s_{-1} = c_{-1}E^{-1}(0) . \tag{13}$$

Note that in the incomplete matrices

$$\begin{bmatrix} * & r_j^* \\ r_j & m_j \end{bmatrix} ; \quad \begin{bmatrix} * & s_{-1}^* \\ s_{-1} & c_{-1} \end{bmatrix} , \tag{14}$$

the smallest matrices which can be placed in the upper left corner so that the matrices obtained are nonnegative are:

$$[E^{-1}(\alpha_j)]^* m_j [E^{-1}(\alpha_j)], \quad [E^{-1}(0)]^* c_{-1}[E^{-1}(0)] .$$

From this and from the convergence of the expansion of the upper diagonal block in (12), the condition 4) obviously follows.

Sufficiency. Let us generate the matrix functions

$$A(z) = \begin{bmatrix} T(z) & [F^*(z)]^{-1} \\ F^{-1}(z) & F^{-1}(z)G(z) \end{bmatrix} , \tag{15}$$

and

$$T(z) := -\frac{t_{-1}}{z} - z \sum_j \frac{t_j}{\alpha_j(z - \alpha_j)} + \hat{c}_0 + \hat{c}_1 z , \tag{16}$$

where

$$t_{-1} := [E^{-1}(0)]^* c_{-1} E^{-1}(0), \quad t_j := [E^{-1}(\alpha_j)]^* m_j E^{-1}(\alpha_j) , \tag{17}$$

$\hat{c}_0 = \hat{c}_0^*$, $\hat{c}_1 \geq 0$ are matrices of order n. The absolute convergence of the expansion (16) follows from 4). If

$$W(z) = \Omega(A(z)) = \|w_{jk}(z)\|_{j,k=1}^2$$

then

$$w_{21}(z) = F(z), \quad w_{22}(z) = G(z) ,$$

and the matrix functions $w_{21}(z)$ and $w_{22}(z)$ are entire. Indeed $w_{11}(z) = F(z)T(z)$. The matrix function $w_{11}(z)$ can have singularities only in points α_j and 0; by making use of (13) we find that $\mathrm{Res}\, w_{11}(z)\big|_{z=\alpha} = t_j F(\alpha_j) = [E^{-1}(\alpha_j)]^* r_j F(\alpha_j) = 0$ and similarly $\mathrm{Res}\, w_{11}(z)\big|_{z=0} = 0$. Put $w_{12}(z) := w_{11}(z)a_{22}(z) - [w_{21}^*(z)]^{-1}$. As above we verify that the matrix function $w_{12}(z)$ is entire. The constructed matrix function $W(z) = \|w_{jk}(z)\|_1^2$ is the desired one.

Note 1. If in (16) we put $\hat{c}_0 = 0$, $\hat{c}_1 = 0$, we get the matrix function $W_0(z)$ which solves Problem 2 with a certain form of asymptotic behavior; all other solutions to Problem 2 are obtained by a leftward multiplication of the matrix function $W_0(z)$ by the block matrix function $V(z)$ of order $2n$ with blocks

$$v_{11} = v_{22} = I_n, \quad v_{12} = \check{c}_0 + \check{c}_1 z, \quad v_{21} = 0 \,,$$

where $\check{c}_0 = \check{c}_0^*$, $\check{c}_1 \geq 0$ are arbitrary matrices of order n.

Note 2. It follows from the above reasoning that, if conditions 1) and 2) of Theorem 4 are fulfilled, and the sets of zeros of the functions $\det F(z)$ and $\det G(z)$ do not intersect, Problem 2 is solvable. The sufficient part of Theorem 3, in particular, follows from this. The sets of zeros of functions $\det F(z)$ and $\det G(z)$ do not intersect if conditions 1) and 2) are fulfilled and rank $m_j = n-$rank $F(\alpha_j)$, rank $c_{-1} = n-$rank $F(0)$. It has been established at the same time that as long as the matrix functions F, G meet the conditions of Theorem 4 (or Theorem 5), the following formula holds

$$[w_{11}(z)i + w_{12}(z)][w_{21}(z)i + w_{22}(z)]^{-1}$$

$$= \alpha + \beta z + \frac{1}{\pi} \int_{-\infty}^{\infty} \left\{ \frac{1}{\lambda - z} - \frac{\lambda}{1+\lambda^2} \right\} [E(\lambda)E^*(\lambda)]^{-1} d\lambda \,, \tag{18}$$

where

$$E(\lambda) = w_{22}(\lambda) + iw_{21}(\lambda) \quad (W(z) = \|w_{jk}(z)\|_{j,k=1}^2) \,.$$

In the real case, the matrix function (18) offers a recipe for the construction of the upper block row of the matrix function $W(z)$. A formula analogous to (18) can be written for various τ (instead of i) out of an open upper half-plane. This leads to recipes for the construction of the matrix function $W(z)$ in the general case.

Theorem 5 *Let F and G be entire matrix functions of order n, with $\det F \not\equiv 0$ and $\det G \not\equiv 0$. The matrix functions F and G form the block row of a certain matrix function of the class B_n if and only if the following conditions hold:*

1) $F^{-1}G \in \hat{C}e_n$;

2) there exists an absolutely convergent expansion of the form

$$[F^*(z)]^{-1}G^{-1}(z) = z\sum_j \frac{d_j}{\gamma_j(z - \gamma_j)} \,,$$

$$\sum_j \gamma_j^{-2}\|d_j\| < \infty \,, \tag{19}$$

where $\{\gamma_j\} = \{\alpha_k\} \cup \{\beta_l\}$; $\{\alpha_k\}$, $\{\beta_l\}$ are the sets of roots of the functions $\det F(z)$ and $\det G(z)$, and d_j are Hermitian matrices of order n.

3) $\det[G(x) + iF(x)] \neq 0$ *for all* $x \in (-\infty, \infty)$.

If conditions 1)–3) are fulfilled, and, besides, F and G are real, there will be a $W \in B_{n;r}$ which has $[F; G]$ as its block row.

The proof of Theorem 5 follows the same path as the proof of Theorem 4, but it requires additional considerations.

The method used in the proof of Theorems 3 and 4 provides an answer to question 3.

References

[1] M. G. Krein. *On the undetermined case of the Sturm-Liouville boundary problem on the interval* $(0, \infty)$. Izv. Akad. Nauk SSSR Ser. Mat. 16 (1952), pp. 293–324 (Russian).

[2] V. P. Potapov. *The multiplicative structure of J-nonextending matrix-functions.* Tr. Mosk. Mat. Ob-va 4 (1955), pp. 25–131 (Russian).

[3] I. C. Gohberg, M. G. Krein. *The theory of Volterra operators in a Hilbert Space its Applications.* Nauka, Moscow, 1967, 508 p.

[4] M. G. Krein. *Fundamental propositions of the representation theory for Hermitian operators with deficiency index* $(m; m)$. Ukrain. Math. J. 1 (1949), N 2, pp. 3–66. English transl. Amer. Math. Soc. (2) 97 (1971), pp. 75–143.

[5] M. G. Krein, S. N. Saakyan. *On some new results in the Theory of Resolvents of Hermitian operators.* Dokl. Adad. Nauk SSSR, 169 (1966) N. 6, pp. 1269–1272.

[6] L. de Branges. *The Expansion Theorem for Hilbert Space of Entire Functions.* Entire Functions and Related Part of Analysis. 1968, AMS.

[7] D. Z. Arov. *Realizations of the Canonical System with the Dissipative Boundary Condition at One End of the Segment according to Dynamic Compliance Coefficients.* Sib. Math. Zhurn. 16 (1975) N 3, pp. 440–463.

[8] M. G. Krein. *On the Theory of Entire Functions of Exponential Type.* Izv. Akad. Nauk SSSR, Ser. Math. 11 (1947), pp. 309–326 (Russian).

[9] A. G. Chackii. *On one of the transformations of Operator Matrices.* Izv. Vysh. Ucheb. Zav. Mathematics 5 (1972), pp. 104–108 (Russian).

Addendum

The results obtained in this article were announced in an article published by the authors in Dokl. Akad. Nauk Ukrainian SSR 1982 Ser. A, N 2, pp. 14–18.

I have taken the liberty of writing this afterword in the hope that the reader may be interested in the opinions and some personal characteristics of V. P. Potapov and M. G. Krein, as well as in the paper itself. Actually, this paper has been familiarly called "the Potapov paper," because of its connection with what Krein described as "the fundamental studies of V. P. Potapov." Krein apparently owed his interest in these subjects to H. Weyl.

Weyl had some special influence on Krein's work in general. Krein said, "Whenever, as a young man, I would ask myself what was the main thing in science, I would think of what Weyl was doing, and that helped me get my bearings." It seems that Krein and Potapov took diametrically opposite views of Weyl's work.[1] Krein believed that Weyl's paper's was the first step towards future sections of harmonic analysis and that such a theory should be developed via the general theory of resolvent matrices, Hermitian and positive Hermitian operators. He even had an epigraph ready for one of the continuations Weyl might have written to the "Potapov paper:" "The longer one keeps silent, the more important the words are which one says afterwards." Weyl published his paper a quarter of a century after his first work on ordinary differential equations. Potapov's assessment was very categorical — "it was a shallow paper." He never changed this view as far as I know, but he applied equally strict criteria to himself. "I was 'arguing' with Weyl for many years, making arduous attempts to develop a theory containing nothing else than analysis rather than some circles or radii. The result was that I realized he was an outstanding mathematician, whereas I was not."

Coming back to the paper mentioned above, I would like to emphasize that there was another meaning behind the phrase "Potapov paper," namely that operators, as distinct from matrix constructions, are cited there. The involvement of operators is sometimes direct, sometimes requires its own "operator ideology." Krein often regretted that, when M. S. Livšic moved to Kharkov, the cooperation between "operator expert" M. S. Livšic and "analysis expert" V. P. Potapov was broken. "Livšic and Potapov are a tremendous power when they are together," he said. Another opportunity that Potapov missed, in Krein's opinion, was his failure to make active use of the theory of Hilbert

---

[1] H. Weyl. Über das Pick-Nevanlinnashe Interpolationsproblem und sein infinitesimales analogon. Ann. of Math. 36 (1935), pp. 230–254.

spaces of entire functions, developed by Louis de Branges.[2] Although both of them regarded this theory as a remarkable achievement in analytical thought, they had in mind completely different ways for its development.

I.E.Ovcharenko
Krasnoznamennaja str. 7/9, apt.28
Kharkov 310002, Ukraine

---

[2]I think he planned to make use of it, but he died before he could do it. It is a sad circumstance that, shortly before his death, Potapov decided that his security in his position depended on whether he managed to provide an adequate pure complex analysis proof for de Branges's theorem on the uniqueness of the reconstruction of a canonical system of second-order differential equations from its monodromy matrix.

Operator Theory:
Advances and Applications, Vol. 72
© 1994 Birkhäuser Verlag Basel

SOME PROPERTIES OF LINEAR-FRACTIONAL
TRANSFORMATIONS AND THE HARMONIC MEAN
OF MATRIX FUNCTIONS

A. A. Nudel'man

The given linear fractional transformation maps a class of pairs of analytic matrix functions onto a certain set. The conditions under which this set is convex are found. Some new properties of the harmonic mean of matrix functions are obtained.

1. The following construction arises in numerous problems. Using the problems data we may find the 2×2 matrix functions

$$A(z) = \left[\begin{array}{cc} a_{11}(z) & a_{12}(z) \\ a_{21}(z) & a_{22}(z) \end{array} \right] \qquad (z \in G),$$

and then the solutions of the problem are represented by the formula

$$W(z) = \frac{a_{11}(z)w(z) + a_{12}(z)}{a_{21}(z)w(z) + a_{22}(z)},$$

where $w(z)$ runs over the given class F of analytic functions. The following are examples of such problems: the moment problem on the axis, or on a semiaxis, or on a finite interval; classical interpolation problems in various classes of analytic functions; the extension problems for Hermitian-positive functions; the description of all spectral functions of a boundary problem or a string, etc. (see, for example, [5], [2]).

The following classes of functions which are analytic in G are used in the above problems:

(1°) $R = \{w(z)|\ \text{Im } w(z) \geq 0\}, \quad G = \{z|\ \text{Im} > 0\}.$

(2°) $S = \{w(z)|w(z) \in R, \quad w(x) \geq 0, \ x < 0\}.$

It is known (see, for example, [5]) that $w(z) \in S$ iff $w(z) \in R$ and $zw(z) \in R$.

(3°) $B = \{w(z)|\ |w(z)| \leq 1\}, \quad G = \{z|\ |z| \leq 1\}.$

2. V. P. Potapov and his students have used the J-theory developed by him to investigate the matrix-valued generalization of classical interpolation problems and the moment
problem (see, for example, [3] and [4]). In matrix-valued cases the entries a_{ij} of the matrix
$A(z)$ are $m \times m$ matrix functions, and the following classes $\hat{F}$ of pairs of analytic matrix
functions $p(z), q(z)$ are used instead of the classes F:

$$
\begin{aligned}
1° \quad \hat{R} &= \{(p(z), q(z))|\tfrac{p(z)^*q(z)-q(z)^*p(z)}{i} \geq 0\}, \quad G = \{z|\ \text{Im } z \geq 0\}. \\
2° \quad \hat{S} &= \{(p(z), q(z))|(p(z), q(z)) \in \hat{R}, \ (zp(z), q(z)) \in \hat{R}\}. \\
3° \quad \hat{B} &= \{(p(z), q(z))|p(z)^*p(z) - q(z)^*q(z) \leq 0\}, \quad G = \{z|\ |z| \leq 1\}.
\end{aligned} \tag{1}
$$

Each pair $(p(z), q(z))$ belonging to $\hat{F}$ must be nondegenerate. This means that for
every fixed $z \in G$ the equalities

$$
p(z)f = 0, \ q(z)f = 0 \quad (f \in \mathbf{C}^m)
$$

are possible simultaneously only for $f = 0$.

It is easy to see that if $q(z)$ is invertible, then the matrix function $w(z) = p(z)q(z)^{-1}$
has the property

$$
\begin{aligned}
1° \quad &\tfrac{w(z)-w(z)^*}{i} \geq 0, & &\text{if } (p(z), q(z)) \in \hat{R}. \\
2° \quad &\tfrac{w(z)-w(z)^*}{i} \geq 0, \ \tfrac{zw(z)-(zw(z))^*}{i} \geq 0, & &\text{if } (p(z), q(z)) \in \hat{S}. \\
3° \quad &I - w(z)^*w(z) \geq 0, & &\text{if } (p(z), q(z)) \in \hat{B}.
\end{aligned} \tag{2}
$$

We use the symbols R, S, and B respectively (with the common notation F) for the classes
of matrix functions which possess these properties. It should be noted that a matrix function $q(z)$ which belongs to the pair $(p(z), q(z)) \in B$ must be invertible. All solutions of
matrix-valued generalizations of classical problems are represented by the linear fractional
transformation

$$
W(z) = (a_{11}(z)p(z) + a_{12}(z)q(z))(a_{21}(z)p(z) + a_{22}(z)q(z))^{-1}
$$

of some classes $\hat{F}$. V. P. Potapov and his students investigated analytic and J-properties of the corresponding matrix function $A(z)$.

The main object of consideration in this paper is a linear fractional transformation

$$L((p(z), q(z)) = (a_{11}(z)p(z) + a_{12}(z)q(z))(a_{21}(z)p(z) + a_{22}(z)q(z))^{-1} \quad (z \in G)$$

of a class $\hat{F}$ with negligible requirements for the block matrix $A(z) = [a_{ij}(z)]$; neither analyticity of $A(z)$, nor J-properties being supposed.

Let's denote the domain of transformation L by D_L,

$$D_L = \{(p(z), q(z))|(a_{21}(z)p(z) + a_{22}(z)q(z))^{-1} \text{ exists for all } z \in G\}.$$

V. P. Potapov proved [6] that D_L is nonempty iff

$$\text{rank}(a_{21}(z), a_{22}(z)) = m \text{ for all } z \in G.$$

It is easy to see that if $r(z)$ is invertible for every $z \in G$, then

$$L(p(z)r(z), q(z)r(z)) = L(p(z), q(z)).$$

Moreover, if the pair $(p(z), q(z))$ satisfies one of the conditions (1), then $(p(z)r(z), q(z)r(z))$ satisfies the same conditions. Therefore we shall say that a pair $(p(z), q(z))$ belongs to the class $\hat{R}$, $\hat{S}$, or $\hat{B}$, if it is nondegenerate, satisfies the corresponding condition (1), and there exists an invertible matrix function $r(z)(z \in G)$ such that matrix functions $p(z)r(z)$, $q(z)r(z)$ become analytic in G.

Let's denote by $L(D_L \cap \hat{F})$ the set of all matrix functions $W(z)$ which admit a representation

$$W(z) = L(p(z), q(z)),$$

where $(p(z), q(z)) \in D_L \cap \hat{F}$.

3. We shall study the following

PROBLEM. *Find the necessary and sufficient condition for the set $L(D_L \cap \hat{F})$ to be convex.*

This condition will give some *a priori* information about blocks of the matrix $A(z)$ in the classical problems since the set of solutions of such problem is convex.

There are some useful relationships. If

$$W_0(z) = L(p_0(z), q_0(z)), W_1(z) = L(p_1(z), q_1(z)),$$

then for

$$W_t(z) = (1 - t)W_0(z) + tW_1(z), \quad 0 < t < 1,$$

we get

$$W_t(z) = L(p_t(z), q_t(z)),$$

where for $0 < t < 1$,

$$
\begin{aligned}
p_t(z) &= (1 - t)p_0(z)(a_{21}(z)p_0(z) + a_{22}(z)q_0(z))^{-1} \\
&+ tp_1(z)(a_{21}(z)p_1(z) + a_{22}(z)q_1(z))^{-1}, \\
q_t(z) &= (1 - t)q_0(z)(a_{21}(z)p_0(z) + a_{22}(z)q_0(z))^{-1} \\
&+ tq_1(z)(a_{21}(z)p_1(z) + a_{22}(z)q_1(z))^{-1}.
\end{aligned}
\tag{3}
$$

Indeed, it can be verified immediately that

$$a_{21}(z)p_t(z) + a_{22}(z)q_t(z) = I,$$

so $(p_t(z), q_t(z)) \in D_L$ for all t. Now we may obtain

$$W_t(z) = L(p_t(z), q_t(z)) = a_{11}(z)p_t(z) + a_{12}(z)q_t(z) = (1 - t)W_0(z) + tW_1(z).$$

If $a_{21}(z)$ is invertible we may introduce the function

$$u(z) = a_{21}(z)^{-1}a_{22}(z),$$

and if, in addition, $q_0(z)$ and $q_1(z)$ are invertible, then we may take

$$
\begin{aligned}
p_t(z) &= (1 - t)w_0(z)(w_0(z) + u(z))^{-1} + tw_1(z)(w_1(z) + u(z))^{-1} \\
q_t(z) &= (1 - t)(w_0(z) + u(z))^{-1} + t(w_1(z) + u(z))^{-1}
\end{aligned}
$$

where $w_j(z) = p_j(z)q_j(z)^{-1}$, $j = 0, 1$. In this case

$$p_t(z) + u(z)q_t(z) = I,$$

and if $q_t(z)$ is invertible, we obtain

$$p_t(z)q_t(z)^{-1} = q_t(z)^{-1} - u(z).$$

Now for the midpoint $W(z)$ of the segment with the endpoints $W_0(z)$ and $W_1(z)$ we have $W(z) = L(w(z), I)$, where

$$
\begin{aligned}
w(z) \;&=\; p_{\frac{1}{2}}(z)q_{\frac{1}{2}}(z)^{-1} \\
&=\; 2((w_0(z) + u(z))^{-1} + (w_1(z) + u(z))^{-1})^{-1} - u(z).
\end{aligned}
$$

We see that for the midpoint of the segment the matrix function $w(z) + u(z)$ is the harmonic mean of the matrix functions $w_0(z) + u(z)$ and $w_1(z) + u(z)$:

$$w(z) + u(z) = H(w_0(z) + u(z), w_1(z) + u(z)),$$

where $H(A, B) = 2(A^{-1} + B^{-1})^{-1}$. In other terms the harmonic mean is the doubled parallel sum of the given matrices (see, for example, [1]).

4. Considering the sets $L(D_L \cap \hat{R})$ and $L(D_L \cap \hat{S})$, we suppose that

$$(0, I) \in D_L, \qquad (I, 0) \in D_L,$$

in other words, the blocks $a_{21}(z), a_{22}(z)$ of the matrix $A(z)$ are invertible. This requirement is valid for classical problems. Recall that $u(z) = a_{21}(z)^{-1}a_{22}(z)$.

THEOREM 1. *Let $(0, I) \in D_L$, $(I, 0) \in D_L$. The set $L(D_L \cap \hat{R})$ is convex if and only if $u(z) \in R$.*

PROOF. Let $L(D_L \cap \hat{R})$ be a convex set. Since $(0, I) \in D_L$, $(I, 0) \in D_L$, then the matrix functions

$$W_0(z) = L(0, I), \quad W_1(z) = L(I, 0)$$

belong to the set $L(D_L \cap \hat{R})$, and hence the midpoint

$$W(z) = L(p_{\frac{1}{2}}(z), q_{\frac{1}{2}}(z))$$

belongs to $L(D \cap \hat{R})$ too, so $(p_{\frac{1}{2}}(z), q_{\frac{1}{2}}(z)) \in \hat{R}$ for $p_0 = 0$, $q_0 = I$, $p_1 = I$, $q_1 = 0$.

Using (3) we obtain $p_{\frac{1}{2}}(z) = \frac{1}{2}a_{21}(z)^{-1}, q_{\frac{1}{2}}(z) = \frac{1}{2}a_{22}(z)^{-1}$. Hence $p_{\frac{1}{2}}(z)q_{\frac{1}{2}}(z)^{-1} \in R$, that is $a_{21}(z)^{-1}a_{22}(z) = u(z) \in R$.

Conversely, let $u(z) \in R$, $W_j(z) = L(p_j(z), q_j(z))$, $(p_j(z), q_j(z)) \in \hat{R}$, $j = 0, 1$, and $W(z) = \frac{W_0(z) + W_1(z)}{2} = L(p(z), q(z))$, where according to (3),

$$
\begin{aligned}
p(z) &= p_0(z)A_0(z) + p_1(z)A_1(z) \\
q(z) &= q_0(z)A_0(z) + q_1(z)A_1(z) \\
A_j(z) &= (p_j(z) + u(z)q_j(z))^{-1}, \quad j = 0, 1.
\end{aligned}
$$

To prove that $W(z) \in L(D_L \cap \hat{F})$, we must verify that $(p(z), q(z)) \in \hat{R}$. This is easily deduced from the following identity.

$$
\begin{aligned}
&\frac{q(z)^* p(z) - p(z)^* q(z)}{i} \\
&= 2A_0(z)^* \left(\frac{q_0(z)^* p_0(z) - p_0(z)^* q_0(z)}{i} \right) A_0(z) \\
&+ 2A_1(z)^* \left(\frac{q_1(z)^* p_1(z) - p_1(z)^* q_1(z)}{i} \right) A_1(z) \\
&+ (q_0(z)A_0(z) - q_1 A_1(z))^* \left(\frac{u(z) - u(z)^*}{i} \right) (q_0(z)A_0(z) - q_1(z)A_1(z)),
\end{aligned}
$$

since each term of the right-hand side is positive semidefinite. In turn this identity follows from the next ones:

$$
\begin{aligned}
q(z)^* p(z) - p(z)^* q(z) &= A_0(z)^*(q_0(z)^* p_0(z) - p_0(z)^* q_0(z))A_0(z) \\
&+ A_0(z)^*(q_0(z)^* p_1(z) - p_1(z)^* q_0(z))A_1(z) \\
&+ A_1(z)^*(q_1(z)^* p_0(z) - p_0(z)^* q_1(z))A_0(z) \\
&+ A_1(z)^*(q_1(z)^* p_1(z) - p_1(z)^* q_1(z))A_1(z), \quad (4)
\end{aligned}
$$

and

$$
\begin{aligned}
&(q_0(z)A_0(z) - q_1(z)A_1(z))^*(u(z) - u(z)^*)(q_0(z)A_0(z) - q_1(z)A_1(z)) \\
&= -A_0(z)^*(q_0(z)^* p_0(z) - p_0(z)^* q_0(z))A_0(z) \\
&+ A_0(z)^*(q_0(z)^* p_1(z) - p_0(z)^* q_1(z))A_1(z) \\
&+ A_1(z)^*(q_1(z)^* p_0(z) - p_1(z)^* q_0(z))A_0(z) \\
&- A_1(z)^*(q_1(z)^* p_1(z) - p_1(z)^* q_1(z))A_1(z).
\end{aligned}
$$

The right-hand sides of the two latter identities are obtained by direct computation, and for the second one it must be noted that $u(z)q_j(z)A_j(z) = I - p_j(z)A_j(z)$, $j = 0, 1$.

THEOREM 2. *The set $L(D_L \cap \hat{S})$ is convex if and only if $u(z) \in S$.*

PROOF is analogous to that of Theorem 1. If the set $L(D_L \cap \hat{S})$ is convex, then for $W_0(z) = L(0, I), W_1(z) = L(I, 0)$ the midpoint $W(z) = L(p_{\frac{1}{2}}(z), q_{\frac{1}{2}}(z))$ belongs to $L(D_L \cap \hat{S})$, hence $(p_{\frac{1}{2}}(z), q_{\frac{1}{2}}(z)) \in \hat{S}$, that is, $p_{\frac{1}{2}}(z)q_{\frac{1}{2}}(z)^{-1} = u(z) \in S$. Conversely, if $u(z) \in S$, then the convexity of $L(D_L \cap \hat{S})$ follows from the identities (4) and

$$
\frac{q(z)^*(zp(z)) - (zp(z))^*q(z)}{i}
$$
$$
= 2A_0(z)^* \left(\frac{q_0(z)^*(zp_0(z)) - (zp_0(z))^*q_0(z)}{i} \right) A_0(z)
$$
$$
+ 2A_1(z)^* \left(\frac{q_1(z)^*(zp_1(z)) - (zp_1(z))^*q_1(z)}{i} \right) A_1(z)
$$
$$
+ (q_0(z)A_0(z) - q_1(z)A_1(z))^* \left(\frac{(zu(z) - (zu(z))^*}{i} \right) (q_0(z)A_0(z) - q_1(z)A_1(z)). \quad (5)
$$

THEOREM 3. *Let $(I, I) \in D_L, (-I, I) \in D_L$. The set $L(D_L \cap \hat{B})$ is convex if and only if $v(z) = a_{22}(z)^{-1}a_{21}(z) \in \hat{B}$.*

PROOF. The assumptions $(I, I) \in D_L, (-I, I) \in D_L$, mean that the matrix functions $a_{21}(z) + a_{22}(z)$ and $-a_{21}(z) + a_{22}(z)$ are invertible. It is easy to see that $(p(z), q(z)) \in \hat{R}$ iff $(p(z) - iq(z), p(z) + iq(z)) \in \hat{B}$. Since

$$
L(p(z) - iq(z), p(z) + iq(z))
$$
$$
= ((a_{11}(z) + a_{12}(z))p(z) + i(a_{12}(z) - a_{11}(z))q(z))((a_{21}(z)
$$
$$
+ a_{22}(z))p(z) + i(a_{22}(z) - a_{21}(z))q(z))^{-1},
$$

where $(p(z), q(z)) \in \hat{R}$, then according to Theorem 1 the set $L(D_L \cap \hat{B})$ is convex if and only if

$$
i(a_{21}(z) + a_{22}(z))^{-1}(a_{22}(z) - a_{21}(z)) \in R, \quad (6)
$$

or taking into account (2) if and only if

$$
(a_{22}(z) - a_{21}(z))(a_{21}(z) + a_{22}(z))^* + (a_{21}(z) + a_{22}(z))(a_{22}(z) - a_{21}(z))^* \geq 0,
$$

or

$$
a_{22}(z)a_{22}(z)^* - a_{21}(z)a_{21}(z)^* \geq 0. \quad (7)
$$

The block $a_{22}(z)$ is invertible for all $z \in G$. Indeed, if for some $z_0 \in G$ and $f \in \mathbf{C}^m$ we have $f^* a_{22}(z_0) = 0$, then from (7) we deduce that

$$-f^* a_{21}(z_0) a_{21}(z_0)^* f \geq 0,$$

that is, $f^* a_{21}(z_0) = 0$, and so $f^*(a_{21}(z_0) + a_{22}(z_0)) = 0$. But the matrix $a_{21}(z_0) + a_{22}(z_0)$ is invertible, so $f = 0$. Now for $v(z) = a_{22}^{-1}(z) a_{11}(z)$, we obtain that (6) is equivalent to

$$i(I + v(z))^{-1}(I - v(z)) \in R,$$

and hence $v(z)$ is analytic in G. Finally, (7) is equivalent to

$$I - v(z)v(z)^* \geq 0,$$

that is, $v(z) \in B$.

5. The next two theorems do not deal with the convexity of the set $L(D_L \cap \hat{F})$, but nevertheless, they are closely related to Theorems 1 and 3.

THEOREM 4. *The set $\hat{R}$ is a subset of the set D_L if and only if the inequality*

$$\frac{u(z) - u(z)^*}{i} > 0, \quad \text{for all } z \in G \tag{8}$$

is valid.

PROOF. Suppose (8) is true, but for some $z_0 \in G$ and for some pair $(p(z), q(z)) \in \hat{R}$ the matrix $p(z_0) + u(z_0)q(z_0)$ is not invertible. Then there exists $f \in \mathbf{C}^m$, $f \neq 0$ such that $(p(z_0) + u(z_0)q(z_0))f = 0$. Let's put $p(z_0)f = g$ and $q(z_0)f = h$, hence $u(z_0)h = -g$. It is easy to see that $h \neq 0$. Indeed, if $h = 0$, then $g = 0$, that is $p(z_0)f = 0$, $q(z_0)f = 0$. Since the pair $(p(z), q(z))$ is nondegenerate, the last two equalities are true only for $f = 0$, but this is impossible. Now from (8) it follows that

$$h^* \left(\frac{u(z_0) - u(z_0)^*}{i} \right) h > 0,$$

or

$$\frac{h^* g - g^* h}{i} < 0. \tag{9}$$

On the other hand, $(p(z), q(z)) \in R$, so we obtain from (1) that

$$f^* \left(\frac{q(z_0)^* p(z_0) - p(z_0)^* q(z_0)}{i} \right) f \geq 0,$$

or

$$\frac{h^*g - g^*h}{i} \geq 0.$$

The last inequality contradicts (9). The sufficiency of the condition (8) is proved.

To prove the necessity we suppose that $p(z) + u(z)q(z)$ $(z \in G)$ is invertible for all $(p(z), q(z)) \in \hat{R}$, but (8) is false. Then there exist $z_0 \in G$, $f \in C^m$, $f \neq 0$ such that

$$f^* \left(\frac{u(z_0) - u(z_0)^*}{i} \right) f \leq 0. \tag{10}$$

Let's put $u(z_0)f = -g$, hence $\frac{f^*g - g^*f}{i} \geq 0$.

(a) If $g^*f \neq 0$, we put $p = gg^*$, $q = (g^*f)I$. The pair (p, q) belongs to $\hat{R}$ since it is nondegenerate and

$$\frac{q^*p - p^*q}{i} = \frac{((f^*g)gg^* - gg^*(g^*f))}{i} = (gg^*)\frac{f^*g - g^*f}{i} \geq 0.$$

The matrix $p + u(z_0)q$ is not invertible because

$$(p + u(z_0)q)f = g(g^*f) + u(z_0)f(g^*f) = (g + u(z_0)f)g^*f = 0,$$

which contradicts our assumption.

(b) If $g^*f = 0$, we put $p = fg^* + gf^*$, $q = (f^*f)I$. This pair also belongs to $\hat{R}$ because it is nondegenerate and $q^*p - p^*q = 0$. And the matrix $p + u(z_0)q$ is not invertible again since

$$(p + u(z_0)q)f = g(f^*f) + u(z_0)f(f^*f) = (g + u(z_0)f)(f^*f) = 0.$$

The proof is complete.

The next theorem may be proved similarly to Theorem 4.

THEOREM 5. *The set $\hat{B}$ is a subset of the set D_L if and only if the inequality*

$$I - v(z)v(z)^* > 0, \text{ for all } z \in G$$

is valid.

6. For the classical problems in the class $\hat{R}$ there is a set of remarkable solutions, so-called canonical solutions. These solutions correspond to $W(z) = L(p, q)$ where (p, q) is a

pair of constant matrices such that $q^*p - p^*q = 0$. In the scalar case it means that $w = pq^{-1}$ is a real constant.

THEOREM 6. *If*

$$\frac{u(z) - u(z)^*}{i} > 0, \text{ for all } z \in G,$$

then the canonical function $W(z) = L(p, q)(p^*q = q^*p)$ *is an extreme point of the convex set* $L(\hat{R})$.

PROOF. Let $W(z) = L(p, q)(p^*q = q^*p)$ be a canonical function represented in the form

$$W(z) = \frac{W_0(z) + W_1(z)}{2},$$

where $W_j(z) = L(p_j(z), q_j(z)), (p_j(z), q_j(z)) \in \hat{R}, \quad j = 0, 1$.

We may put

$$\begin{aligned}
p &= p_0(z)A_0(z) + p_1(z)A_1(z), \\
q &= q_0(z)A_0(z) + q_1(z)A_1(z) \\
A_j(z) &= (p_j(z) + u(z)p_j(z))^{-1}, \quad j = 0, 1.
\end{aligned} \tag{11}$$

It follows from the condition $p^*q = q^*p$ that the left-hand side of the identity (4) is equal to 0. Since each term of the right-hand side is nonnegative semidefinite, they are equal to 0, too. This is possible only if

$$\begin{aligned}
q_0(z)^*p_0(z) &= p_0(z)^*q_0(z), \\
q_1(z)^*p_1(z) &= p_1(z)^*q_1(z), \\
q_0(z)A_0(z) &= q_1(z)A_1(z).
\end{aligned}$$

We obtain from this equality the next one:

$$p_0(z)A_0(z) = I - u(z)q_0(z)A_0(z) = I - u(z)q_1(z)A_1(z) = p_1(z)A_1(z).$$

Now (see (11)),

$$\begin{aligned}
p &= 2p_0(z)A_0(z) = 2p_1(z)A_1(z); \\
q &= 2q_0(z)A_0(z) = 2q_1(z)A_1(z).
\end{aligned}$$

Hence

$$W_{j}(z) = L(p_{j}(z), q_{j}(z)) = L(2p_{j}(z)A_{j}(z), 2q_{j}(z)A_{j}(z))$$
$$= L(p, q) = W(z), \quad j = 0, 1.$$

This proves that $W(z)$ is the extreme point of the set $L(\hat{R})$.

There are two kinds of canonical solutions of problems in the class $\hat{S}$. Canonical solutions of the first kind correspond to $W(z) = L(p, q)$ where matrices p, q are constant, $q^*p = p^*q \geq 0$, and those of the second kind correspond to $W(z) = L(z^{-1}p, q)$, where matrices p, q are constant, $q^*p = p^*q \leq 0$.

THEOREM 7. *Let $u(z) \in S$.*

(i) *If*

$$\frac{u(z) - u(z)^*}{i} > 0, \quad for \; all \quad z \in G,$$

then the canonical function of the first kind $W(z) = L(p, q)$, $p^*q = q^*p \geq 0$ is the extreme point of the convex set $L(\hat{S})$.

(ii) *If*

$$\frac{zu(z) - (zu(z))^*}{i} > 0, \quad for \; all \quad z \in G,$$

then the canonical function of the second kind $W(z) = L(z^{-1}p, q)$, $p^*q = q^*p \leq 0$ is the extreme point of the convex set $L(\hat{S})$.

PROOF is analogous to that of Theorem 6, and it is based on identities (4) and (5) respectively.

7. We can deduce several new properties of the harmonic mean of matrix functions from Theorems 1-3. A simple computation shows that if an invertible matrix function $w(z)$ belongs to class R, then $-w(z)^{-1}$ belongs to R, too. The sum of two invertible matrix functions belonging to R is invertible. Hence, if $w_0(z) \in R$ and $w_1(z) \in R$ are invertible, then their harmonic mean exists and belongs to R as well.

Theorem 1 allows us to obtain a stronger assertion.

If $u(z)$, $w_0(z)$, $w_1(z) \in R$, and $H(w_0(z) + u(z), w_1(z) + u(z))$ exists, then

$$H(w_0(z) + u(z), w_1(z) + u(z)) - u(z) \in R.$$

This assertion may be reformulated in a simpler form if we introduce a partial order in R in such a way: we say that $g(z) \succ h(z)$ if $g(z) - h(z) \in R$.

If $u(z) \succ 0$, $v_0(z) \succ u(z)$, $v_1(z) \succ u(z)$ and $H(v_0(z), v_1(z))$ exists, then

$$H(v_0(z), v_1(z)) \succ u(z).$$

If $g(z) \succ h(z)$ means that $g(z) - h(z) \in S$, we can deduce the same corollary from Theorem 2.

Finally, to obtain a corollary from Theorem 3, for given $\epsilon > 0$, and $h(z)$, we put

$$U(h, \epsilon) = \{g(z) | (g(z) - h(z))^*(g(z) - h(z)) \leq \epsilon^2 I\}$$

(all matrix functions are analytic). The next corollary follows from Theorem 3:

If $h(z)^*h(z) \geq \epsilon^2 I$ and $g_0(z), g_1(z) \in U(h, \epsilon)$, then $H(g_0(z), g_1(z)) \in U(h, \epsilon)$ as well.

Indeed, the matrix function $v(z) = \left(\frac{h(z)}{\epsilon}\right)^{-1}$ belongs to the class B, thus, according to Theorem 3 the linear fractional transformation

$$L(w(z)) = w(z)(v(z)w(z) + I)^{-1}$$

maps B onto a convex set. Since $w_j(z) = \frac{(g_j(z) - h(z))}{\epsilon}$, $j = 0, 1$, belong to B, the matrix function

$$
\begin{aligned}
w_{\frac{1}{2}}(z) &= H(w_0(z) + v(z)^{-1}, w_1(z) + v(z)^{-1}) - v(z)^{-1} \\
&= \frac{H(g_0(z), g_1(z))}{\epsilon} - \frac{h(z)}{\epsilon}
\end{aligned}
$$

also belongs to B. Thus $H(g_0(z), g_1(z)) = h(z) + \epsilon w_{\frac{1}{2}}(z) \in U(h, \epsilon)$.

The above-mentioned properties of the harmonic mean may be reformulated for classes of matrix functions which are obtained from one of classes R, or S, or B by means of a linear fractional transformation. The following theorem may be used for this purpose.

THEOREM 8. If

$$(v_{\frac{1}{2}} + u)^{-1} = \frac{(v_0 + u)^{-1} + (v_1 + u)^{-1}}{2}, \tag{12}$$

and

$$
\begin{aligned}
V_j &= (a_{11}v_j + a_{12})(a_{21}v_j + a_{22})^{-1}, \quad j = 0, 1, \frac{1}{2}, \\
U &= (ub_{11} + b_{21})^{-1}(ub_{12} + b_{22}),
\end{aligned}
$$

where

$$\begin{bmatrix} b_{11} & b_{12} \\ b_{21} & b_{22} \end{bmatrix} \begin{bmatrix} a_{11} & a_{12} \\ a_{21} & a_{22} \end{bmatrix} = \begin{bmatrix} 0 & I \\ I & 0 \end{bmatrix}, \qquad (13)$$

then

$$(V_{\frac{1}{2}} + U)^{-1} = \frac{((V_0 + U)^{-1} + (V_1 + U)^{-1})}{2}.$$

PROOF. Using (13) we observe that

$$\begin{aligned} V_j + U &= (ub_{11} + b_{21})^{-1}((ub_{11} + b_{21})(a_{11}v_j + a_{12}) \\ &\quad + (ub_{12} + b_{22})(a_{21}v_j + a_{22}))(a_{21}v_j + a_{22})^{-1} \\ &= (ub_{11} + b_{21})^{-1}(v_j + u)(a_{21}v_j + a)^{-1}. \end{aligned}$$

Hence,

$$\begin{aligned} (V_j + U)^{-1} &= (a_{21}v_j + a_{22})(v_j + u)^{-1}(ub_{11} + b_{21}) \\ &= (a_{21}(v_j + u) + a_{22} - a_{21}u)(v_j + u)^{-1}(ub_{11} + b_{21}) \\ &= A + B(v_j + u)^{-1}C, \end{aligned}$$

where $A = a_{21}(ub_{11} + b_{21})$, $B = a_{22} - a_{21}u$, $C = ub_{11} + b_{21}$ does not depend on v_j. So, if (12) is true, then

$$\begin{aligned} (V_{\frac{1}{2}} + U)^{-1} &= A + B(v_{\frac{1}{2}} + u)^{-1}C \\ &= A + B((v_0 + u)^{-1} + (v_1 + u)^{-1})\frac{C}{2} \\ &= \frac{((V_0 + U)^{-1} + (V_1 + U)^{-1}}{2}. \end{aligned}$$

References

[1] W. N. Anderson, Jr. and R. J. Duffin, Series and parallel addition of matrices, *J. Math. Anal. Appl.*, 3(1969), 576-594.

[2] I. S. Kac and M. G. Krein, On the spectral function of the string, Appendix 2 to the Russian translation of the book by F. V. Atkinson, Discrete and continuous boundary problems, *Mir*, Moscow, 1968, 648-737. English translation, Amer. Math. Soc. Translation (2) 103(1973), 19-102.

[3] I. V. Kovalishina and V. P. Potapov, Indefinite metric in the Nevanlinna-Pick problem. *Dokl. Akad. Nauk Armjan. SSR*, 59(1974), N1, 17-22. English translation: Collected papers of V. P. Potapov, private translation and edition by T. Ando, 1982, Sapporo, Japan, 33-40.

[4] I. V. Kovalishina, *J*-expansive matrix functions and the classical problem of moments. *Dokl. Akad. Nauk Armjan. SSR*, 60(1975), N1, 3-10.

[5] M. G. Krein and A. A. Nudel'man, The Markov moment problem and extremal problems, Translation Math. Monographs, Vol. 50, Amer. Math. Soc., Providence, Rhode Island, 1977.

[6] V. P. Potapov, Fractional-linear transformations of matrices. *Studies in the Theory of Operators and their Applications,* Kiev, 1979, 75-97. English translation: Collected papers of V. P. Potapov, private translation and edition by T. Ando, 1982, Sapporo, Japan, 41-65.

Odessa Civil Engineering Institute
Didrikhson Street, 4
270029 Odessa
THE UKRAINE

MSC 1991
Primary: 47A56
Secondary: 47D20, 47A63

Operator Theory:
Advances and Applications, Vol. 72
© 1994 Birkhäuser Verlag Basel

MODIFICATION OF V.P. POTAPOV'S
SCHEME IN THE INDEFINITE CASE

A.L. Sakhnovich

The problem to determine integral representations and structure preserving extensions of operators with κ negative eigenvalues is investigated. A modification of the scheme of V.P. Potapov (combined with the method of operator identities due to L.A. Sakhnovich) is proposed. Toeplitz matrices, operators with difference kernels and operators with sum and difference kernels are considered as examples. A generalization of Szegö's limit theorem is formulated.

0   Introduction

The problem of finding integral representations and structure preserving extensions of operators with κ negative eigenvalues is an important problem in interpolation theory. The most investigated case is the case $\kappa = 0$; see [1] and the bibliography there. Various articles are dedicated to the case $\kappa \neq 0$. The integral representations of Toeplitz matrices were introduced in [2]. The continuous analogue of this problem was solved in [3]. These results are connected with conditionally positive definite functions and their generalizations; see [4] – [6]. Further progress in this area was initiated by the well-known article [7] of V.M. Adamjan, D.Z. Arov and M.G. Krein, where the description of the extensions of operators from important classes was given in terms of linear fractional transformations. Finally V.P. Potapov proposed the Method of Matrix Inequalities to solve interpolation problems; see [8], [9] and for a more detailed version [10]. The transfer matrix function, the elements of which define the coefficients of the linear fractional transformation, was constructed by solving certain matrix inequalities. Though V.P. Potapov proposed his method for the case $\kappa = 0$, his ideas proved to be fruitful for $\kappa \neq 0$ also. These ideas were essentially used in [11] – [16]. Unification of V.P. Potapov's ideas with L.A. Sakhnovich's method of the operator identities, c.f. [17] – [19], leads to the interesting works [20], [21] and [1]. Here we generalize results of [12] - [14], taking into account [8], [9], [20] and [21].

1   Preliminaries

1. We begin with the folowing definitions.

Definition 1 $\{H_1, H_2\}$ is the set of all linear bounded operators acting from H_1 to H_2. $(H_1, H_2$ are Hilbert spaces.)

Definition 2 $\mathcal{P}_\kappa$ is the set of all hermitian operators $S \in \{H, H\}$ such that their negative spectrum consists of κ negative eigenvalues.

Definition 3 (See [22].) N_κ is the set of all functions $\varphi(\lambda)$, which are meromorphic on the open upper halfplane $\mathbf{C}_+$ and such that the kernel

$$\nu_\varphi(\lambda, \mu) = (\lambda - \bar{\mu})^{-1}(\varphi(\lambda) - \overline{\varphi(\mu)})$$

has κ negative squares, i.e, for any positive integer n and any $\lambda_1, \cdots, \lambda_n \in \mathbf{C}_+$ the matrix $\{\nu_\varphi(\lambda_k, \lambda_j)\}_{k,j=1}^n$ has at most κ negative eigenvalues and for at least one choice of $n, \lambda_1, \cdots, \lambda_n$ it has exactly κ negative eigenvalues. It will be useful to define φ in the lower half plane by the equality

$$\varphi(\lambda) = \overline{\varphi(\tilde{\lambda})}. \tag{1}$$

We shall describe the integral representations of operators $S \in \mathcal{P}_\kappa$ in terms of functions $\varphi \in N_\kappa$.

Theorem 4 (See [22].) *Every function $\varphi \in N_\kappa$ may be represented in the form*

$$\varphi(\lambda) = \int_{-\infty}^\infty \left(\frac{1}{t - \lambda} + \sum_{j=0}^k K_j(t, \lambda) \right) (1 + t^2)^{\rho_0 + 1} g(t) d\tau(t) +$$
$$+ R_0(\lambda) + \sum_{j=1}^k R_j \left(\frac{1}{\lambda - \alpha_j} \right) + \sum_{j=1}^m [M_j \left(\frac{1}{\lambda - \beta_j} \right) + \overline{M_j \left(\frac{1}{\bar{\lambda} - \beta_j} \right)}], \tag{2}$$

where
a) $k \geq 0, \ m \geq 0, \ \alpha_1, \alpha_2, \cdots, \alpha_k$ are different real numbers, $\beta_1, \beta_2, \cdots, \beta_m$ are different complex numbers with $\mathrm{Im}\beta_j > 0$ and

$$g(t) = \prod_{j=1}^k (1 + t^2)^{\rho_j} / (t - \alpha_j)^{2\rho_j};$$

b) $\tau(t)$ is a bounded, nondecreasing, left continuous function;
c) the functions K_j are given by

$$K_0(t, \lambda) = -(t + \lambda) \sum_{p=1}^{\rho_0 + 1} \frac{(1 + \lambda^2)^{p-1}}{(1 + t^2)^p} \chi_{v_0}.$$

$$K_j(t, \lambda) = \sum_{p=1}^{2\rho_j} \frac{(t - \alpha_j)^{p-1}}{(\lambda - \alpha_j)^p} \chi_{v_j}, \quad 1 \leq j \leq k,$$

where

$$\lambda_{U_j}(t) = \begin{cases} 1, & t \in U_j \\ 0, & t \notin U_j \end{cases}, \quad 0 \le j \le k,$$

and $U_1, U_2, \cdots, U_k$ and U_0 are nonintersecting neighbourhoods of the points $\alpha_1, \alpha_2, \cdots, \alpha_k$ and ∞:

d) $R_j(\lambda)$ and $M_j(\lambda)$ are polynomials, the coefficients of the $R_j(\lambda)$'s are real numbers, the degree of $R_0(\lambda)$ is no more than $2\rho_0 + 1$, the degree of the $R_j(\lambda)$ is no more than $2\rho_j$, $R_j(0) = 0$, $1 \le j \le k$, and $M_j(0) = 0$;

e) the degrees ξ_j of $M_j(\lambda)$'s satisfy the equality

$$\sum_{j=0}^{k} \rho_i + \sum_{j=1}^{m} \xi_j = \kappa. \tag{3}$$

After the choice of the neighbourhoods U_j the representation (2) is uniquely determined.

Let us denote the coefficients of $R_0(\lambda)$ by c_k :

$$R_0(\lambda) = c_{2\rho_0+1}\lambda^{2\rho_0+1} + \cdots + c_0.$$

Remark 5 We have that

$$c'_{2\rho_0+1} = c_{2\rho_0+1} - \int_{U_0} g(t)d\tau(t) \ge 0.$$

2. The operators $A, S \in \{H, H\}$ and $\Phi_1, \Phi_2 \in \{H_0, H\}$ form an S-node (see [19]) if $S = S^*$ and the operator identity

$$AS - SA^* = i(\Phi_1\Phi_2^* + \Phi_2\Phi_1^*) \tag{4}$$

is satisfied. We shall further suppose that $\dim H_0 = 1$, i.e., $H_0 = \mathbb{C}$. Let us fix operators A and Φ_2 and a function $\varphi \in N_\kappa$, and put for $j > 0$,

$$\begin{aligned} \mathcal{F}_j &= -\text{res}_{\lambda = \gamma_j} \ (E - \lambda A)^{-1}\Phi_2 F_j(\lambda)\Phi_2^*(E - \lambda A^*)^{-1}, \\ \hat{\mathcal{F}}_j &= -\text{res}_{\lambda = \gamma_j} \ (E - \lambda A)^{-1}A\Phi_2 F_j(\lambda), \end{aligned} \tag{5}$$

where the quadruple $\{F_j(\lambda), \gamma_j, \mathcal{F}_j, \hat{\mathcal{F}}_j\}$ stands for one of the following quadruples

$$\{K_j(t,\lambda), \alpha_j, \mathcal{K}_j(t), \hat{\mathcal{K}}_j(t)\}, \quad \left\{R_j\left(\frac{1}{\lambda - \alpha_j}\right), \alpha_j, \mathcal{R}_j, \hat{\mathcal{R}}_j\right\},$$

$$\left\{M_j\left(\frac{1}{\lambda - \beta_j}\right), \beta_j, \mathcal{M}_{1j}, \hat{\mathcal{M}}_{1j}\right\}, \quad \left\{\overline{M_j\left(\frac{1}{\lambda - \beta_j}\right)}, \beta_j, \mathcal{M}_{2j}, \hat{\mathcal{M}}_{2j}\right\}.$$

Let us put also

$$\begin{aligned} \mathcal{F}_0 &= \text{res}_{\lambda=0}(A - \lambda E)^{-1}\Phi_2 F_0(1/\lambda)\Phi_2^*(A^* - \lambda E)^{-1}, \\ \hat{\mathcal{F}}_0 &= -\text{res}_{\lambda=0}(A - \lambda E)^{-1}A\Phi_2 F_0(1/\lambda)/\lambda, \quad \text{if } 0 \notin \sigma(A), \end{aligned} \tag{6}$$

$$K_0 = 0, \quad \hat{K}_0(t) = \frac{t}{1+t^2} \chi_{U_0} \Phi_2, \quad R_0 = 0, \quad \hat{R}_0 = -c_0 \Phi_2, \quad \text{if } 0 \in \sigma(A), \qquad (7)$$

where $\sigma(A)$ is spectrum of A.

We shall suppose that A satisfies:

Condition I. $\sigma(A)$ is a finite set of points and $(\sigma(A) \setminus \{0\}) \cap (-\infty, \infty) = \emptyset$.

Definition 6 $\mathcal{E}$ is the class of functions $\varphi \in N_\eta$ such that $\beta_j^{-1} \notin \sigma(A)$, $\bar{\beta}_j^{-1} \notin \sigma(A)$, and if $0 \in \sigma(A)$, then: $\rho_0 = 0$, $c_1' = 0$ and for every $h \in H$,

$$\int_{U_0} | \Phi_2^*(E - tA^*)^{-1}h |^2 (1+t^2)d\tau(t) < \infty. \qquad (8)$$

According to (5)-(8) the following lemma holds true.

Lemma 7 *For* $\varphi \in \mathcal{E}$ *the integrals*

$$\begin{aligned} S_\varphi &= \int_{-\infty}^{\infty} [(E - tA)^{-1} \Phi_2 \Phi_2^* (E - tA^*)^{-1} + \sum_{j=0}^{k} \mathcal{K}_j(t)](1+t^2)^{\rho_0+1} g(t) d\tau(t) \\ &\quad + \sum_{j=0}^{k} \mathcal{R}_j + \sum_{j=1}^{m} (\mathcal{M}_{1j} + \mathcal{M}_{1j}^*) \end{aligned} \qquad (9)$$

and

$$\begin{aligned} i\Phi_\varphi &= \int_{-\infty}^{\infty} [(E - tA)^{-1} A\Phi_2 + \sum_{j=0}^{k} \hat{\mathcal{K}}_j(t)](1+t^2)^{\rho_0+1} g(t) d\tau(t) + \\ &\quad + \sum_{j=0}^{k} \hat{\mathcal{R}}_j + \sum_{j=1}^{m} (\hat{\mathcal{M}}_{1j} + \hat{\mathcal{M}}_{2j}) \end{aligned} \qquad (10)$$

converge in a weak sense and the operators S_φ, Φ_φ *are well defined. The operators* A, S_φ, Φ_φ *and* Φ_2 *form an S-node and satisfy the operator identity*

$$AS_\varphi - S_\varphi A^* = i(\Phi_\varphi \Phi_2^* + \Phi_2 \Phi_\varphi^*). \qquad (11)$$

3. Let us put $\Pi = [\Phi_1, \Phi_2]$, $E_2 = \{\delta_{kj}\}_{k,j=1}^2$, $J = \{1 - \delta_{kj}\}_{k,j=1}^2$, and (supposing $0 \notin \sigma(S)$)

$$\mathfrak{M}(\lambda) = \begin{bmatrix} a(\lambda) & c(\lambda) \\ b(\lambda) & d(\lambda) \end{bmatrix} = E_2 - i\lambda \Pi^*(E - \lambda A^*)^{-1} S^{-1} \Pi J. \qquad (12)$$

Definition 8 $\mathcal{N}(\mathfrak{M})$ is the class of linear fractional transformations

$$\varphi(\lambda) = i[a(\lambda)\psi(\lambda) + ib(\lambda)][c(\lambda)\psi(\lambda) + id(\lambda)]^{-1}, \qquad (13)$$

where $\psi \in N = N_0 \cup \infty$, $c\psi + id \not\equiv 0$. (If $\psi = \infty$, then the condition $c\psi + id \not\equiv 0$ means that $c \neq 0$ and the transformation (13) takes the form $\varphi(\lambda) = ia(\lambda)c(\lambda)^{-1}$.)

It is easy to see that $\varphi(\lambda) \in \mathcal{N}(\mathfrak{M})$ satisfies (1), as $\psi(\lambda) = \psi(\bar{\lambda})$. (The proof requires only the equality (16) below.)

Lemma 9 *Let* A, S, Π *be an S-node and assume that* $S \in \mathcal{P}_\kappa$, $0 \notin \sigma(S)$. *Let* $\varphi \in \mathcal{N}(\mathfrak{M})$. *Then* $\varphi \in N_\eta$, $\eta \leq \kappa$.

Proof. From (4) and (12) it follows that (see [19]):

$$\mathfrak{M}(\mu).J\mathfrak{M}^*(\lambda) = J + i(\lambda - \mu)\amalg^*(E - \mu A^*)^{-1}S^{-1}(E - \lambda A)^{-1}\amalg. \tag{11}$$

According to (14), the equalities

$$\frac{\varphi(\lambda) - \overline{\varphi(\mu)}}{\lambda - \mu} = [1, i\overline{\varphi(\mu)}]\amalg^*(E - \mu A^*)^{-1}S^{-1}(E \quad \lambda A)^{-1}\amalg \begin{bmatrix} 1 \\ i\varphi(\lambda) \end{bmatrix}$$

$$+\frac{i}{\lambda - \mu}[1, i\overline{\varphi(\mu)}]\mathfrak{M}(\mu).J\mathfrak{M}^*(\lambda) \begin{bmatrix} 1 \\ -i\varphi(\lambda) \end{bmatrix}, \tag{15}$$

$$\mathfrak{M}(\lambda).J\mathfrak{M}^*(\lambda) = J \tag{16}$$

are true. By (13) and (16), we get

$$[1, i\overline{\varphi(\mu)}]\mathfrak{M}(\mu).J\mathfrak{M}^*(\lambda) \begin{bmatrix} 1 \\ -i\varphi(\lambda) \end{bmatrix} =$$

$$= i[c(\mu)\psi(\mu) + id(\mu)]^{-1}[\psi(\mu) - \psi(\lambda)][c(\lambda)\psi(\lambda) + id(\lambda)]^{-1}. \tag{17}$$

From (15) and (17) we conclude that, since $w(\lambda) \in N$,

$$\left\{ \frac{\varphi(\lambda_k) - \overline{\varphi(\lambda_l)}}{\lambda_k - \bar{\lambda}_l} \right\}_{k,l=1}^n \geq$$

$$\left\{ [1, i\overline{\varphi(\lambda_l)}]\amalg^*(E - \lambda_l A^*)^{-1}S^{-1}(E - \lambda_k A)^{-1}\amalg \begin{bmatrix} 1 \\ -i\varphi(\lambda_k) \end{bmatrix} \right\}_{k,l=1}^n. \tag{18}$$

As $S^{-1} \in \mathcal{P}_\kappa$, the statement of the lemma now follows from (18).

2 Basic propositions

1. From now on it will be supposed that A, S and $\amalg = [\Phi_1, \Phi_2]$ form an S-node, $S \in P$, and that $0 \notin \sigma(S)$. Analogously to the case $\kappa = 0$, cf. [20], we say that S_φ, defined by (9), is an integral representation of the operator S, if

$$S = S_\varphi \quad . \quad \varphi \in N_\kappa \quad . \quad \varphi \in \mathcal{E}. \tag{19}$$

We shall investigate the interpolation problem of describing the integral representation (19) in terms of the linear fractional transformation (13). The following simple lemma will help us to modify V.P. Potapov's scheme for the case $\kappa \neq 0$.

Lemma 10 Let an operator $L = \{L_{kl}\}_{k,l=1}^2 \in \{H_1 \dotplus H_2, H_1 \dotplus H_2\}$ be given, and suppose that $L_{11} \in \mathcal{P}_\kappa$ and $0 \notin \sigma(L_{11})$. Then the inequality

$$L_{22} - L_{21}L_{11}^{-1}L_{12} \geq 0 \tag{20}$$

is a necessary and sufficient condition for $L \in \mathcal{P}_\kappa$.

2. We use Lemma 10 in the proof of the following theorem.

Theorem 11 *If (19) is valid, then φ has the properties*

$$\varphi + \alpha \in \mathcal{N}(\mathfrak{M}), \quad \Phi_1 = \Phi_\varphi + i\alpha\Phi_2, \tag{21}$$

for some real number α.

Proof. Taking into account the formulas (2) and (10) and putting

$$B_\varphi(\lambda) = (E - \lambda A)^{-1}[\Phi_\varphi - i\varphi(\lambda)\Phi_2], \tag{22}$$

we obtain the representation

$$
\begin{aligned}
iB_\varphi(\lambda) = & \\
& \int_{-\infty}^{\infty}[\gamma(t,\lambda) - \sum_{j=1}^{k} \operatorname{res}_{z=\alpha_j}\gamma(z,\lambda)K_j(t,z) - \gamma_0(t,\lambda)]\Phi_2(1+t^2)^{p_0+1}g(t)d\tau(t) \\
& -[\sum_{j=1}^{k}\operatorname{res}_{z=\alpha_j}\gamma(z,\lambda)R_j\left(\frac{1}{z-\alpha_j}\right) + \sum_{j=1}^{m}[\operatorname{res}_{z=\beta_j}\gamma(z,\lambda)M_j\left(\frac{1}{z-\beta_j}\right) \\
& + \operatorname{res}_{z=\beta_j}\gamma(z,\lambda)M_j\left(\frac{1}{\bar{z}-\beta_j}\right)] + \gamma_1(t,\lambda)]\Phi_2,
\end{aligned}
\tag{23}
$$

where $\gamma(z,\lambda) = (E - zA)^{-1}/(z-\lambda)$,

$$
\begin{aligned}
\gamma_0 &= \begin{cases} \operatorname{res}_{z=0}(1-z\lambda)^{-1}(A-zE)^{-1}K_0(t,z^{-1}), & \text{if } 0 \notin \sigma(A), \\ 0, & \text{if } 0 \in \sigma(A), \end{cases} \\
\gamma_1 &= \begin{cases} \operatorname{res}_{z=0}(1-z\lambda)^{-1}(A-zE)^{-1}R_0(z^{-1}), & \text{if } 0 \notin \sigma(A), \\ 0, & \text{if } 0 \in \sigma(A). \end{cases}
\end{aligned}
\tag{24}
$$

(Here we use the identity

$$(E - \lambda A)^{-1}[A(E - zA)^{-1} + \frac{1}{z-\lambda}E] = (E - zA)^{-1}/(z-\lambda)$$

and the equality $c_1' = 0$, which is valid by Definition 6 if $\varphi \in \mathcal{E}$ and $0 \in \sigma(A)$.)

3. Potapov's Main Matrices have the form

$$
L_\varphi = \begin{bmatrix} S_\varphi & B_\varphi(\lambda) \\ B_\varphi(\lambda)^* & \frac{\varphi(\lambda) - \overline{\varphi(\lambda)}}{\lambda - \bar{\lambda}} \end{bmatrix}, \quad
L = \begin{bmatrix} S & B(\lambda) \\ B(\lambda)^* & \frac{\varphi(\lambda) - \overline{\varphi(\lambda)}}{\lambda - \bar{\lambda}} \end{bmatrix},
\tag{25}
$$

where

$$B(\lambda) = (E - \lambda A)^{-1}[\Phi_1 - i\varphi(\lambda)\Phi_2].$$

Taking into account (2), (9) and (23)-(25), it is easy to see that $L_\varphi \in P_\eta$, $\eta \le \kappa$. According to (19), $S_\varphi \in P_\kappa$. Hence $L_\varphi \in P_\kappa$. Using (25) and Lemma 10 we get

$$\frac{\varphi(\lambda) - \overline{\varphi(\lambda)}}{\lambda - \bar{\lambda}} \ge B_\varphi(\lambda)^* S_\varphi^{-1} B_\varphi(\lambda). \tag{26}$$

From (22) and (26) it follows that

$$\frac{i}{\lambda - \bar\lambda}[\overline{i\varphi(\lambda)}, 1]\left\{J + i(\lambda - \bar\lambda)J\Pi_\varphi^*(E - \bar\lambda A^*)^{-1}S_\varphi^{-1}(E - \lambda A)^{-1}\Pi_\varphi J\right\}\begin{bmatrix} -i\varphi(\lambda) \\ 1 \end{bmatrix} \geq 0,$$
(27)

where $\Pi_\varphi = [\Phi_\varphi, \Phi_2,]$. Let us now compare formulas (4) and (11), taking into account (19). We obtain the equality $(\Phi_1 - \Phi_\varphi)\Phi_2^* + \Phi_2(\Phi_1 - \Phi_\varphi)^* = 0$. Hence (see Lemma 1.2 in [20]) for some $\alpha = \bar\alpha$,

$$\Phi_1 = \Phi_\varphi + i\alpha\Phi_2, \quad \Pi_\varphi\begin{bmatrix} 1 \\ -i\varphi(\lambda) \end{bmatrix} = \Pi\begin{bmatrix} 1 \\ -i(\varphi(\lambda) + \alpha) \end{bmatrix}.$$
(28)

By (14), (19), (27) and (28) we have that

$$(i/(\lambda - \bar\lambda))[i(\overline{\varphi(\lambda)} + \alpha), 1]J\mathfrak{M}(\bar\lambda)J\mathfrak{M}^*(\bar\lambda)J\begin{bmatrix} -i(\varphi(\lambda) + \alpha) \\ 1 \end{bmatrix} \geq 0.$$
(29)

Now formulas (16) and (29) mean that the vector

$$\begin{bmatrix} P(\lambda) \\ Q(\lambda) \end{bmatrix} = \mathfrak{M}^{-1}(\lambda)\begin{bmatrix} -i(\varphi(\lambda) + \alpha) \\ 1 \end{bmatrix}$$
(30)

generates a J-nonnegative pair in $\mathbf{C}_+$: $P(\lambda)\overline{Q(\lambda)} + Q(\lambda)\overline{P(\lambda)} \geq 0$. So the representation

$$\varphi(\lambda) + \alpha = i[a(\lambda)P(\lambda) + b(\lambda)Q(\lambda)][c(\lambda)P(\lambda) + d(\lambda)Q(\lambda)]^{-1},$$

that follows from (30), is equivalent to $\varphi(\lambda) + \alpha \in \mathcal{N}(\mathfrak{M})$. The theorem is proved.

Remark 12 If $\kappa = 0$, then $L_\varphi \in \mathcal{P}_0$, i.e., $L_\varphi \geq 0$. The inequality $L_\varphi \geq 0$ introduced by V.P. Potapov was fundamental for Potapov's approach to interpolation problems. It was named the Main Matrix Inequality. In the case $\kappa \neq 0$ we use inequality (26).

4. Before formulating the next theorem we introduce several conditions.

Condition II. $\sigma(A) \cap \sigma(A^*) = \emptyset$.

Condition III. $\sigma(A) \cap \sigma(A^*) = \{0\}$, $\text{Ker} A = \{0\}$, and for every $f \in H \setminus \{0\}$, there exist $\epsilon > 0$, $\delta > 0$ and $h \in H$, such that

$$|(\lambda(E - \lambda A)^{-1}f, h)| \neq 0(1)$$
(31)

in the domain $|\lambda| \geq \epsilon$, $\lambda \in \mathcal{D}_\delta$, where

$$\mathcal{D}_\delta = \{\lambda : 0 < \delta < |\arg \lambda| < \pi - \delta\}.$$

Condition IV. There is a sequence $\{\lambda_k\}$ such that

$$\lim_{k \to \infty} \varphi(\lambda_k)/\lambda_k = 0, \quad \lambda_k \in \mathcal{D}_\delta, \quad |\lambda_k| \to \infty.$$
(32)

Let us put also

$$B_T(\lambda) = [SA^* + \imath B(\lambda)\Phi_2^*](E - \lambda A^*)^{-1},$$
$$B_{T\varphi}(\lambda) = [S_\varphi A^* + \imath B_\varphi(\lambda)\Phi_2^*](E - \lambda A^*)^{-1}. \tag{33}$$

We recall that

$$B(\lambda) = (E - \lambda A)^{-1}[\Phi_1 - \imath\varphi(\lambda)\Phi_2].$$

For convenience we shall write Φ_1, Φ_2, B instead of $\Phi_1 1$, $\Phi_2 1$, $B1$.

Condition V. For every $f \in H$ and $\delta > 0$ we have the equality

$$|(B(\lambda), f)| = 0(1), \quad \lambda \in \mathcal{D}_\delta, \quad |\lambda| \to \infty,$$

and for every $f \in H$ there are a sequence $\{\mu_k\}$ and $\delta > 0$ such that

$$\varlimsup_{k \to \infty} |\mu_k(B_T(\mu_k)f, f)| < \infty, \quad \mu_k \in \mathcal{D}_\delta, \quad |\mu_k| \to \infty. \tag{34}$$

Theorem 13 *Let* $\varphi \in \mathcal{N}(\mathfrak{M})$ *be given by (13) with* $c(\lambda)\psi(\lambda) + \imath d(\lambda) \neq 0$, *when* $\lambda^{-1} \in \sigma(A)$. *Suppose that either*
a) the operator A *satisfies Conditions I and II, or*
b) the operator A *satisfies Conditions I and III, and the Conditions IV and V are valid.*
Then (19) is true.

Proof. Case a). By Lemma 9, $\varphi \in N_\eta$. $\eta \leq \kappa$. As $c(\lambda)\psi(\lambda) + \imath d(\lambda) \neq 0$ when $\lambda^{-1} \in \sigma(A)$, we have that $\jmath_\imath^{-1}$, $\bar{\jmath}_\imath^{-1} \notin \sigma(A)$. By Condition II, $0 \notin \sigma(A)$. Hence $\varphi \in \mathcal{E}$, and, by Lemma 7. the operators S_φ, Φ_φ exist. If $\lambda \notin \{\beta_j\} \cup \{\bar{\beta}\} \cup (-\infty, \infty)$, then, according to (23), the function $B_\varphi(\lambda)$ is well-defined. On the other hand, taking into account (4) and (12). we obtain the identity

$$(E - \lambda A)^{-1}\Pi J\mathfrak{M}(\lambda) = S(E - \lambda A^*)^{-1}S^{-1}\Pi J. \tag{35}$$

By (13). (33) and (35) we get that

$$B(\lambda) = S(L - \lambda A^*)^{-1}S^{-1}\Pi J \begin{bmatrix} \psi(\lambda) \\ \imath \end{bmatrix} (c(\lambda)\psi(\lambda) + \imath d(\lambda))^{-1}.$$

Hence, according to Condition II, the function $B(\lambda)$ is defined also for λ such that $\lambda^{-1} \in \sigma(A)$. Thus the function

$$B(\lambda) - B_\varphi(\lambda) = (E - \lambda A)^{-1}(\Phi_1 - \Phi_\varphi)$$

is defined for $\lambda^{-1} \in \sigma(A)$. So $(E - \lambda A)^{-1}(\Phi_1 - \Phi_\varphi)$ is an entire function. As $0 \notin \sigma(A)$, we obtain that

$$\Phi_1 = \Phi_\varphi. \tag{36}$$

By Condition II. the operator identity

$$A(S - S_\varphi) - (S - S_\varphi)A^* = 0 \tag{37}$$

has a single solution $S = S_\varphi$. (Equation (37) follows from (4), (11) and (36).) From $S_\varphi = S \in \mathcal{P}_\kappa$ we conclude that $\eta = \kappa$, i.e., $\varphi \in N_\kappa$. Formula (19) is proved.

5. The case b) is more difficult, but it includes several important examples, when H is infinite dimensional.

Potapov's Transformed Main Matrix L_T (see [10], [23], [24], [20] and [1]) may be introduced by the equalities.

$$G = \begin{bmatrix} E & 0 \\ iA(E - \bar{\lambda}A)^{-1} & (E - \bar{\lambda}A)^{-1}\Phi_2 \end{bmatrix}, \quad L_T = GLG^*. \tag{38}$$

By (4), (25), (33) and (38) it is true that

$$L_T = \begin{bmatrix} S & -iB_T(\lambda) \\ iB_T(\lambda)^* & C_T(\lambda,\lambda) \end{bmatrix}, \quad C_T(\lambda,\mu) = \frac{B_T(\lambda) - B_T(\mu)^*}{\lambda - \bar{\mu}}. \tag{39}$$

Taking into account (4), (15), (17), (33) and (39), we obtain

$$C_T(\lambda,\mu) =$$
$$(E - \bar{\mu}A)^{-1}[iA, \Phi_2] \begin{bmatrix} S \\ B(\mu)^* \end{bmatrix} S^{-1}[S, B(\lambda)] \begin{bmatrix} -iA^* \\ \Phi_2^* \end{bmatrix} (E - \lambda A^*)^{-1}$$
$$+ (E - \bar{\mu}A)^{-1}\Phi_2\{[c(\mu)\psi(\mu) + id(\mu)]^*\}^{-1}\frac{\psi(\lambda) - \bar{\psi}(\mu)}{\lambda - \bar{\mu}}[c(\lambda)\psi(\lambda) + id(\lambda)]^{-1}\Phi_2^*(E - \lambda A^*)^{-1}. \tag{40}$$

Hence $B_T(\lambda) = B_T(\bar{\lambda})^*$ and for every $f \in H$,

$$(B_T(\lambda)f, f) \in N_\eta, \quad \eta \le \kappa. \tag{41}$$

By Lemma 9, $\varphi \in N_{\eta_1}(\eta_1 \le \kappa)$. Formula (41) will help us to prove that $\varphi \in \mathcal{E}$, i.e., that the inequality (8) is satisfied. Indeed, from (2) and (41) we get

$$(B_T(\lambda)f, f) = \varphi_0(\lambda) + R_0(\lambda) + 0(|\lambda|^{-1}), \tag{42}$$

where

$$\varphi_0(\lambda) = (1 + \lambda^2)^{p_0+1}\int_{U_0}(t - \lambda)^{-1}g_f(t)d\tau_f(t). \tag{43}$$

It is not difficult to draw from the existence of the irreducible representation of $\varphi \in N_\kappa$ (see [22]) and formulas (34), (42) and (43) the conclusion that $p_0 = 0$, $c_1' = 0$ and

$$\varphi_0(\lambda) + R_0(\lambda) = c_0 + \int_{U_0}\frac{1 + t\lambda}{t - \lambda}g_f(t)d\tau_f(t). \tag{44}$$

According to (34), (42) and (44) it is true that

$$\int_{U_0}(1 + t^2)g_f(t)d\tau(t) < \infty. \tag{45}$$

By the Stieltjes-Perron theorem, taking into account (2). (33)and (44) we get for $\lambda = x + iy$,
$w_f = \int_0^t (1 + u^2) g(u) \chi_{t_0}(u) d\tau_f(u)$ and $[t_1,\ t_2] \subset U_0$.

$$[w_f(t_2 + 0) + w_f(t_2 - 0) - w_f(t_1 + 0) - w_f(t_1 - 0)] =$$
$$= (2/\pi) \lim_{y \to 0} \int_{t_1}^{t_2} \text{Im}(B_T(\lambda)f, f) dr =$$
$$= (-i/\pi) \lim_{y \to 0} \{ \int_{t_1}^{t_2} [\varphi(\lambda) \Phi_2^*(E - \lambda A^*)^{-1} f \overline{\Phi_2^*(E - \lambda A^*)^{-1} f}$$
$$- \overline{\varphi(\lambda)} \Phi_2^*(E - \lambda A^*)^{-1} f \overline{\Phi_2^*(E - \lambda A^*)^{-1} f}] dr \}.$$
(16)

As $\varphi \in N_m$, formula (32) means that in the representation (2) of $\varphi(\lambda)$ we have that

$$\rho_0 = 0 \ , \quad c_1' = 0. \tag{17}$$

From (46)and (47) we conclude that for sufficiently large $|t_1|, |t_2| (t_1 t_2 > 0)$, where $\tau(t)$ has no jumps:

$$w_f(t_2) - w_f(t_1) =$$
$$= (-i/2\pi) \lim_{y \to 0} \int_{t_1}^{t_2} [\varphi(\lambda) - \overline{\varphi(\lambda)}] \cdot |\Phi_2^*(E - x A^*)^{-1} f|^2 dr$$
$$= \int_{t_1}^{t_2} (1 + u^2) g(u) |\Phi_2^*(E - u A^*)^{-1} f|^2 d\tau(u). \tag{18}$$

According to (45). (47)and (48), $\varphi \in \mathcal{E}$ and the operators S_φ, Φ_φ, $B_\varphi(\lambda)$ are defined. As in case a) it may be shown that $(E - \lambda A)^{-1}(\Phi_1 - \Phi_\varphi)$ is an entire function. By (23) and Condition V it is true that for $h \in H$,

$$|(B_\varphi(\lambda), h)| = 0(1) \ , \quad |(B(\lambda), h)| = 0(1) \ , \lambda \in D_\lambda, |\lambda| \to \infty.$$

As $B(\lambda) - B_\varphi(\lambda) = (E - \lambda A)^{-1}(\Phi_1 - \Phi_\varphi)$, we get that

$$|((E - \lambda A)^{-1}(\Phi_1 - \Phi_\varphi), h)| = 0(1).$$

Hence

$$|\lambda((E - \lambda A)^{-1} A(\Phi_1 - \Phi_\varphi), h)| = 0(1) \ , \quad \lambda \in \mathcal{D}_\mathcal{E}. \ h \in H. \tag{19}$$

From (49) and Condition III it follows that $A(\Phi_1 - \Phi_\varphi) = 0$, i.e.,

$$\Phi_1 = \Phi_\varphi. \tag{50}$$

Taking into account (9), (23) and (33), we conclude that

$$B_{T_\varphi}(\lambda) =$$
$$= \int_{-\infty}^\infty \left[\frac{i(t)}{t - \lambda} + \sum_{j=1}^k \text{res}_{z = \alpha_j} \frac{v(z)}{\lambda - z} K_j(t, z) \right] g(t)(1 + \tau^2) d\tau(t) + \sum_{j=1}^k \frac{i(-)}{\lambda - z} R_j \left(\frac{1}{\tau - \alpha_j} \right)$$
$$+ \sum_{j=1}^m \left[\text{res}_{z = \beta_j} \frac{v(z)}{\lambda - z} M_j \left(\frac{1}{z - \beta_j} \right) + \text{res}_{z = \bar{\beta}_j} \frac{v(z)}{\lambda - z} M_j \left(\frac{1}{z - \bar{\beta}_j} \right) \right], \tag{51}$$

where $v(z) = (E - zA)^{-1}\Phi_2\Phi_2^*(E - zA^*)^{-1}$. Thus $B_{T\varphi}(\lambda)$ is defined for $\lambda \notin \{\beta_j\} \cup \{\bar{\beta}_j\} \cup (-\infty, \infty)$, and

$$| (B_{T\varphi}(\lambda)f, f) |= 0(| \lambda |^{-1}) , \quad \lambda \in \mathcal{D}_\delta, | \lambda | \to \infty. \tag{52}$$

From (4) and (33) we obtain

$$\imath(E - \lambda A)^{-1}\Phi_1\Phi_2^*(E - \lambda A^*)^{-1} + \imath(E - \lambda A)^{-1}\Phi_2\Phi_1^*(E - \lambda A^*)^{-1} + \\ +SA^*(E - \lambda A^*)^{-1} = [(E - \lambda A)^{-1} - E]S/\lambda, \tag{53}$$

$$B_T(\lambda) = \lambda^{-1}[(E - \lambda A)^{-1} - E]S - \imath(E - \lambda A)^{-1}\Phi_2 B^*(\bar{\lambda}). \tag{54}$$

So when $\lambda^{-1} \in \sigma(A^*)$, the function $B_T(\lambda)$ is defined. According to (33) and (50),

$$B_T(\lambda) - B_{T\varphi}(\lambda) = (S - S_\varphi)A^*(E - \lambda A^*)^{-1}.$$

Hence $(S - S_\varphi)A^*(E - \lambda A^*)^{-1}$ is an entire function. By (34), (42), (44) and (45) we have that

$$| (B_T(\lambda)f, f) |= 0(| \lambda |^{-1}) , \quad \lambda \in \mathcal{D}_\delta, | \lambda | \to \infty. \tag{55}$$

From (52), (55) we get

$$| \lambda((S - S_\varphi)A^*(E - \lambda A^*)^{-1}f, f) |= 0(1) , \quad \lambda \in \mathcal{D}_\delta, | \lambda | \to \infty, f \in H. \tag{56}$$

Taking into account (56), we easily conclude from Condition III that $S = S_\varphi$. It follows, that $\eta_1 = \kappa$, i.e., $\varphi \in N_\kappa$. Representation (19) is proved.

3 Extensions of the operator S.

1. Let the Hilbert space $\hat{H}$ and the operators $\hat{A} \in \{\hat{H}, \hat{H}\}$, $\hat{\Phi}_2 \in \{C, \hat{H}\}$ satisfy the relations

$$\hat{H} \supset H, \quad P\hat{A} = P\hat{A}P, \quad P\hat{A}h = Ah \ (h \in H), \quad P\hat{\Phi}_2 = \Phi_2, \tag{57}$$

where Pf is orthogonal projection of $f \in \hat{H}$ on H.

Definition 14 An operator $\check{S} \in \{\hat{H}, \hat{H}\}$ is called an extension of the operator S, if

$$P\check{S}h = Sh \ (h \in H), \quad \check{A}\check{S} - \check{S}\check{A}^* = \imath[\check{\Phi}_1\check{\Phi}_2^* + \check{\Phi}_2\check{\Phi}_1^*]. \tag{58}$$

The operators, sets, etc., corresponding to the node $\check{A}$, $\check{S}$, $\check{\Pi}$, instead of the node A, S, Π, will be marked with a " ˇ ", for example, $\check{\mathcal{E}}$, $\check{\mathfrak{M}}$, $\check{S}_\rho$ and so on.

There are important examples when $\mathcal{E} = \hat{\mathcal{E}}$:

Theorem 15 Let $\hat{A}$ satisfy Condition I, let φ satisfy (19), and let $\mathcal{E} = \hat{\mathcal{E}}$. Then the operator $\check{S}_\varphi$ is an extension of the operator S, and $\check{S}_\varphi \in P_\kappa$.

Theorem 15 follows from (9), (57) and Lemma 7.

2. We prove the following theorem.

Theorem 16 *Suppose that the operator $\tilde{S} \in \mathcal{P}_\kappa$ is an extension of the operator S, $0 \notin \sigma(\tilde{S})$, $\sigma(A) \subset \sigma(\check{A})$ and that $\mid \tilde{c}(\lambda) \mid + \mid \check{d}(\lambda) \mid \neq 0$, when $\lambda^{-1} \in \sigma(\check{A})$. Suppose also that the sets $\{A, S, \Pi, \varphi\}$ and $\{\check{A}, \check{S}, \check{\Pi}, \check{\varphi}\}$ satisfy the conditions of Theorem 13 (for every $\varphi \in \mathcal{N}(\mathfrak{M})$, $\check{\varphi} \in \mathcal{N}(\check{\mathfrak{M}})$ such that $c(\lambda)\psi(\lambda) + id(\lambda) \neq 0$ $(\lambda^{-1} \in \sigma(A))$ and $\tilde{c}(\lambda)\check{\psi}(\lambda) + i\check{d}(\lambda) \neq 0 (\lambda^{-1} \in \sigma(\check{A})))$. Then there is a function $\varphi \in \mathcal{N}(\mathfrak{M})$, which defines the extension $\tilde{S}$: $\tilde{S} = \tilde{S}_\varphi$.*

Proof According to the factorization theorem in [25]

$$\check{\mathfrak{M}}(\lambda) = \mathfrak{M}(\lambda)V(\lambda), \quad V(\lambda) = E - i\lambda\check{\Pi}^*\check{S}^{-1}T_{22}^{-1}(E - \lambda A_{22}^*)^{-1}\mathcal{P}_2\check{S}^{-1}\check{\Pi}J, \qquad (59)$$

where $\mathcal{P}_2 = E - \mathcal{P}$, the operators $A_{22}, T_{22} \in \{\check{H} \ominus H, \check{H} \ominus H\}$ are defined by $A_{22}f = \mathcal{P}_2\check{A}f$, $T_{22}f = \mathcal{P}_2\check{T}f$, $f \in \check{H} \ominus H$, and $\check{T} = \check{S}^{-1}$. From (59) it follows that

$$i(\bar{\lambda} - \lambda)[V^*(\lambda)JV(\lambda) - J] \geq 0$$

and the linear fractional transformation

$$\psi(\lambda) = i[v_{11}(\lambda)\check{\psi}(\lambda) + iv_{12}(\lambda)][v_{21}(\lambda)\check{\psi}(\lambda) + iv_{22}(\lambda)]^{-1}, \qquad (60)$$

where $V = \{v_{kj}\}_{k,j=1}^2$, transforms $\check{\psi} \in N$ into $\psi \in N_0$. Let us choose a $\check{\psi}$, which satisfies the inequality

$$\tilde{c}(\lambda)\check{\psi}(\lambda) + i\check{d}(\lambda) \neq 0, \quad \lambda^{-1} \in \sigma(\check{A}). \qquad (61)$$

Then, by (59) and (60), we have that

$$\hat{c}\check{\psi} + i\check{d} = [c\psi + id](-i)[v_{21}\psi + v_{22}]. \qquad (62)$$

Taking into account (61) and (62) we get: $c(\lambda)\psi(\lambda) + id(\lambda) \neq 0$ for $\lambda^{-1} \in \sigma(A) \subset \sigma(\check{A})$. According to (13), (59) and (60), $\varphi = \check{\varphi}$. Hence $\tilde{S} = \tilde{S}_{\check{\varphi}} = \tilde{S}_\varphi$.

4   Examples

1. We shall first consider an interesting and well-known example of Toeplitz matrices. For several other examples with $\dim H < \infty$, see [20] and [21].

Example 17 Let us put, cf. [26],

$$S = S_n = \{s_{j-k}\}_{k,j=1}^n \text{ with } s_p = \check{s}_{-p},$$

$$A = \{A_{kj}\}_{k,j=1}^n, \text{ where } a_{kj} = \begin{cases} 0, & \text{if } k < j, \\ i/2, & \text{if } k = j, \\ i, & \text{if } k > j, \end{cases}$$

$$\Phi_1 = \begin{bmatrix} s_0/2 \\ s_0/2 + s_{-1} \\ \cdots \\ s_0/2 + \sum_{p=1}^{n-1} s_{-p} \end{bmatrix}, \quad \Phi_2 = \begin{bmatrix} 1 \\ 1 \\ \cdots \\ 1 \end{bmatrix}. \qquad (63)$$

Here A satisfies Conditions I and II, and

$$\sigma(A) = \{i/2\}, \quad c(-2i) = -(S^{-1}\Phi_2)_1, \quad d(-2i) = 1 - (S^{-1}\Phi_1)_1. \tag{64}$$

We assume that $0 \notin \sigma(S)$, so that

$$| c(-2i) | + | d(-2i) | \neq 0.$$

(Indeed, if $c(-2i) = d(-2i) = 0$, then the solution T of the equation $TA - A^*T = iS^{-1}\Pi J\Pi^*S^{-1}$ is not invertible, because its last column is equal to zero; see [26]. This contradicts the equality $T = S^{-1}$.) For $\check{S} = S_r (r > n)$ we have, according to (64), that $\mathcal{E} = \check{\mathcal{E}}$. So Theorems 11, 13 and 16 are valid. It is easy to see that

$$(E - \lambda A)^{-1}\Phi_2 = (1 - \frac{i}{2}\lambda)^{-1}
\begin{bmatrix}
1 \\
q(\lambda) \\
\cdots \\
q(\lambda)^{n-1}
\end{bmatrix}, \quad q(\lambda) = \frac{2 + i\lambda}{2 - i\lambda}.$$

Hence the transition from the representations (9) and (19) to the representation

$$s_r = 4[\mathrm{res}_{\lambda=-2i} q(\lambda)^{-p}\varphi(\lambda)/(\lambda^2 + 4) + \mathrm{res}_{\lambda=2i} q(\lambda)^{-p}\varphi(\lambda)/(\lambda^2 + 4)]. \tag{65}$$

is possible; see [11]. From Theorems 11, 13, 15 and 16 we obtain the following result.

Corollary 18 (See ([11] and [12].) Let the S-node (63) be given. Suppose that $S \in P_\kappa$ and $0 \notin \sigma(S)$. Then for every $r > n$ and $\psi(\lambda) \in N$ such that

$$c(-2i)\psi(-2i) + d(-2i) \neq 0,$$

formulas (13) and (65) define extensions $S_r = \{s_{j-k}\}_{k,j=1}^i \in P_\kappa$ of the operator S. Moreover, every extension $S_r \in P_\kappa$ can be represented in this way.

2. Formula (59) and Corollary 18 allow one to prove a generalization for $\kappa \neq 0$ of the Szegö limit theorem. Let us consider a set $\{s_p\}_{p=-\alpha}^\infty$ $(s_{-p} = s_p)$ such that $S_r \in P_\kappa$ and $\Delta_r = \det S_r \neq 0$ for all $r > r_0$. Then the function $\varphi(z) = (s_0/2) + \sum_{p=1}^\infty s_{-p}z^p$ is defined in the unit circle. Its poles will be designated by $z_1, z_2, \cdots, z_m$ and their multiplicities by $\xi_1, \xi_2, \cdots, \xi_m$, respectively.

Theorem 19 (See [12].) For $f(u) = \varphi(e^{iu}) + \overline{\varphi(e^{iu})}$ it is true that

$$\lim_{r \to \infty} \frac{\Delta_{r+1}}{\Delta_r} = \prod_{j=1}^m | z_j |^{-2\xi_j} \exp\left\{\frac{1}{2\pi}\int_{-\pi}^\pi \ln f(u)du\right\}, \quad \sum_{j=1}^m \xi_j \leq \kappa.$$

It would be interesting to generalize the results of [27] connected with the Szegö limit theorem and the maximal jump of the spectral function to the case $\kappa \geq 0$.

3. In this article we essentially use the representations of the functions $\varphi \in N_\kappa$, cf. [22]. When $\dim H_0 > 1$ matrix functions $\varphi \in N_\kappa$ appear. Though the theory of matrix functions $\varphi \in N_\kappa$ is less promoted, the work [28] allows one to obtain some interesting results. Another approach to this problem was given in [11]; see also [14]. Representation (65) was used there to describe the extensions of the block Toeplitz matrices $S \in \mathcal{P}_\kappa$.

4. Before considering other examples we formulate a remark.

Remark 20 For $\varphi \in \mathcal{N}(\mathfrak{M})$, we have that

$$\frac{\varphi(\lambda) - \overline{\varphi(\lambda)}}{\lambda - \bar{\lambda}} \geq B^*(\lambda)S^{-1}B(\lambda), \quad C_T(\lambda, \lambda) \geq B_T^*(\lambda)S^{-1}B_T(\lambda). \tag{66}$$

Proof. The first formula in (66) follows from (18). Hence, by Lemma 10, $L \in \mathcal{P}_\kappa$. Taking into account (38) and (39), we obtain $L_T \in \mathcal{P}_\kappa$, and the second formula in (66) is true.

5. We now come to the second example.

Example 21 Let us introduce now the S-node, where $H = L^2(0, \ell)$ and S is an operator with difference kernel:

$$S = S_\ell = \frac{d}{dx} \int_0^\ell s(x - t) \cdot dt \in \mathcal{P}_\kappa,$$

$$A = i \int_0^x \cdot dt, \tag{67}$$

$$s(x) = -\overline{s(-x)} \in L^2(0, \ell),$$

$$\Phi_1 1 = s(x), \quad \Phi_2 1 = 1.$$

It is easy to see that $c\psi + id \not\equiv 0$ for $\psi \in N_0$ and that

$$(E - \lambda A)^{-1} f = f + i\lambda \int_0^x e^{i\lambda(x-t)} f(t)dt,$$

$$\Phi_2^*(E - \lambda A)^{-1} f = \int_0^\ell e^{i\lambda(x-t)} f(t)dt, \tag{68}$$

$$(E - \lambda A^*)^{-1} f = f - i\lambda \int_x^\ell e^{i\lambda(x-t)} f(t)dt.$$

From (68) it follows that Conditions I and III are satisfied. As $(E - \lambda A)^{-1} = e^{i\lambda x}$, we obtain for every $h \in H$ the equality

$$\lim_{|\lambda| \to \infty} ((E - \lambda A)^{-1}\Phi_2, h) / \parallel (E - \lambda A)^{-1}\Phi_2 \parallel = 0. \tag{69}$$

If for $\varphi \in \mathcal{N}(\mathfrak{M})$ formula (32) is not true, then by (33), (68) and (69) there exist $\epsilon > 0$ and $M > 0$ such that

$$B(\lambda)^* S^{-1} B(\lambda) > \epsilon \mid \varphi(\lambda) \mid^2 \parallel (E - \lambda A)^{-1}\Phi_2 \parallel^2, \quad \lambda \in \mathcal{D}_\delta, \quad \text{Im } \lambda < -M. \tag{70}$$

According to (66) and (70), Condition IV is satisfied and (32) is true. As

$$\| (E - \lambda A)^{-1} \| = 0(1), \quad \lambda \in \mathcal{D}_\delta, \ \mathrm{Im} \ \lambda > 0,$$

$$\| (E - \lambda A^*)^{-1} \| = 0(1), \quad \lambda \in \mathcal{D}_\varepsilon, \ \mathrm{Im} \ \lambda < 0,$$

by (33) and (68), we get the representation

$$B(\lambda) = w(\lambda)(E - \lambda A)^{-1}\Phi_2 + \hat{w}(\lambda), w(\lambda) \in C, \ \| \hat{w}(\lambda) \| = 0(1), \ \lambda \in \mathcal{D}_\varepsilon. \tag{71}$$

From the first formula in (66), (69) and (71) it follows that $\| B(\lambda) \| = 0(1)$. According to (54), we have the equality

$$B_T(\lambda)f = w_1(\lambda)(E - \lambda A)^{-1}\Phi_2 + \hat{w}_1(\lambda), \tag{72}$$

where

$$\| \hat{w}_1(\lambda) \| = 0(| \lambda |^{-1}), \quad \lambda \in \mathcal{D}_\delta, \ \mathrm{Im}\lambda > 0, \ | \lambda | \rightarrow \infty.$$

From the second formula in (66), (39) and (72) we get

$$\| B_T(\lambda)f \| = 0(| \lambda |^{-1}), \quad \lambda \in \mathcal{D}_\varepsilon, \ \mathrm{Im}\lambda > 0, \ | \lambda | \rightarrow \infty.$$

Thus Condition V is satisfied. As I.V. Mihailova observed, here we have that $\mathcal{E} = \dot{\mathcal{E}}$. This easily follows from (68). Hence Theorems 11, 13 and 16 entail the following result.

Corollary 22 *Let the S-node (67) be given. Suppose that $S \in \mathcal{P}_\kappa$ and that $0 \notin \sigma(S)$. Then for every $\psi(\lambda) \in N$, $\tilde{\ell} > \ell$, formulas (9) and (13) define extensions $S_{\tilde{i}} = \dot{S}_\varrho \in \mathcal{P}_\kappa$ of the operator S. Moreover, every extension $S_{\tilde{i}} \in \mathcal{P}_\kappa$ can be represented in this way.*

Corollary 22 is closely connected with the results of [13]. The particular case

$$Sf = \alpha f + \int_0^\ell k(x - t)f(t)dt \tag{73}$$

was investigated in [15]. But the extensions of S, constructed in [15], do not necessarily have the form (73).

6. We conclude the paper with the following example.

Example 23 Putting $H = L^2(0, \ell)$,

$$\begin{aligned}
S &= \tfrac{d^2}{dx^2} \int_0^\ell [s(x - t) + s(x + t)] \cdot dt \in \{H, H\}, \\
A &= \int_0^x (t - x) \cdot dt, \\
\Phi_1 1 &= 2is(x), \quad \Phi_2 1 = 1, \\
s(x) &= \overline{s(x)} = s(-x) \quad (0 < x < \ell),
\end{aligned} \tag{74}$$

where $S \in \mathcal{P}_\kappa$, $0 \notin \sigma(S)$, we can prove quite analogously to the case of the S-node (67) that for $\varphi \in \mathcal{N}(\mathfrak{M})$ the Conditions I and III-V are satisfied.

References

[1] Ivanchenko T.S., Sakhnovich L.A., An operator approach to the Potapov's scheme of investigating the interpolation problems, this issue.

[2] Iohvidov I.S., Krein M.G., Spectral theory of operators in spaces with an indefinite metric II, Trudy Moscov. Mat. Obsc., 8 (1959), 413-496.

[3] Pluscheva V.I. (Gorbachuk V.I.), On the integral representation of Hermitian-indefinite kernels with a finite number of negative squares, Dokl. Akad. Nauk SSSR, 145:3 (1962), 534-537.

[4] Horn, R., The theory of infinitely divisable matrices and kernels, Trans. Amer. Math. Soc., 136 (1969), 269-286.

[5] Horn, R., Infinitely divisable positive definite sequences. Trans. Amer. Math. Soc., 136 (1969), 287-303.

[6] Khelifati S., Integral representations of the conditionally positive definite funtions, Dep. UkrNIINTI 29.01.1986 n.338.

[7] Adamjan V.M., Arov D.Z., Krein M.G., Analytic properties of the Schmidt pairs of a Hankel operator and the generalised problem of Schur-Takagi, Matem.Sb., 86:1 (1971), 34-75.

[8] Potapov V.P., The main facts of the theory of J-nonstretching analytical matrix functions, Abstracts of the reports of the Allunion conference on the theory of functions of a complex variable, Kharkov (1971), 179-181.

[9] Kovalishina I.V., Potapov V.P., Integral representation of Hermitian-positive functions, Dep. VINITI 19.06.1981 n. 2984.

[10] Kovalishina I.V., The analytical theory of a certain class of interpolation problems, Izv. Akad. Nauk. SSSR, Ser. Mat., 47:3 (1983), 445-498.

[11] Sakhnovich A.L., On the continuation of block Toeplitz matrices, Funkcional. Anal. (Uljanovsk), n. 14 (1980), 116-127.

[12] Sakhnovich A.L., On the continuation of Toeplitz matrices, Dokl. Akad. Nauk. UkrSSR, Ser. A., n. 7 (1981), 19-24.

[13] Sakhnovich A.L., On the continuation of Toeplitz matrices and their continuous analogues, Abstract of thesis, Kharkov, 1982.

[14] Nudel'man A.A., On a generalization of the classical interpolation problems, Dokl. Akad. Nauk. SSSR, 256:4 (1981), 790-793.

[15] Krein M.G., Langer H., Continuous analogues of orthogonal polynomials on the unit circle with respect to an indefinite weight and related continuation problems, Dokl. Akad. Nauk SSSR, 358:3 (1981), 537-541.

[16] Krein M.G., Langer H., On some continuation problems, which are closely related to the theory of operators in spaces Π_κ, IV, J. of Operator theory, 13 (1985), 299-417.

[17] Sakhnovich L.A., On the similarity of operators, Dokl. Akad. Nauk SSSR, 200:3 (1971), 541-544.

[18] Sakhnovich L.A., Equations with a difference kernel on a finite interval, Russ. Math. Surv., 35:4 (1980), 81-152.

[19] Sakhnovich L.A., Factorisation problems and operator identities, Russ. Math. Surv., 41:1 (1986).

[20] Ivanchenko T.S., Sakhnovich L.A., An operator approach to the investigation of interpolation problems, Dep. UkrNIINTI 8.04.1985 n. 701-85.

[21] Ivanchenko T.S., Sakhnovich L.A., An operator approach to V.P. Potapov's scheme of solving interpolation problems, Ukr. Mat. J., 39:5 (1987).

[22] Krein M.G., Langer H., On some continuation problems, which are closely related to the theory of operators in spaces Π_κ, I, Math. Nachr., 77 (1977), 187-236.

[23] Katznelson V.E., Continuous analogues of the Hamburger-Nevanlinna theorem, Teor. funkc., funkcional. anal. i prilozen. (Kharkov), n. 36 (1981), 31-48.

[24] Katznelson V.E., Methods of the J-theory in the continuous interpolation problems of analysis I, Dep. VINITI 11.01.1983 n. 171-83.

[25] Sakhnovich L.A., On the factorisation of the transfer matrix function, Dokl. Akad. Nauk SSSR, 226:4 (1976), 781-784.

[26] Sakhnovich A.L., On a method of the inversion of Toeplitz matrices, Mat. Issl. (Kishinev), 8:4 (1973), 180-186.

[27] Sakhnovich A.L., On a class of extremal problems, Izv. Akad. Nauk. SSSR, 51:2 (1987), 436-443.

[28] Krein M.G., Langer H., Some propositions on analytic matrix functions related to the theory of operators in the space Π_κ, Acta Sci. Math. (Szeged), 43 (1981), 181-205.

270111 Odessa,
pr. Dobrovolskogo 154,
kv. 199,
Ukraine.

Operator Theory:
Advances and Applications, Vol. 72
© 1994 Birkhäuser Verlag Basel

Inverse problems for equations systems

L.A Sakhnovich

An inverse problem? Havn't I done everything there? (A joke by V.P. Potapov)

1   Introduction

In inverse problems it is necessary to restore a system from some information about this system (scattering data, spectral data). In the present case the following problems have to be solved:

I. To prove that there exists a system corresponding to the data (existence theorem).

II. To prove that there is a unique system corresponding to the data.

III. To find the method of constructing this system by the data.

V.P. Potapov [6] studied systems of the form

$$(1.1) \qquad \frac{dW}{dx} = i\lambda J\, H(x)\, W(x), \qquad\qquad 0 \le x \le \ell,$$

where the matrices J and $H(x)$ are such that

$$(1.2) \qquad\qquad J = J^*, \qquad J^2 = I_n, \qquad H(x) \ge 0,$$

where I_n denotes the $n \times n$ identity matrix.

With the normalization condition
$$(1.3) \qquad\qquad W(0, \lambda) = I_n$$
the matrix $W(\lambda) = W(\ell, \lambda)$ is the matrix of monodromy of the system (1.1).
In this article the following inverse problem will be studied: to recover the system (1.1), i.e. to find $H(x)$, from the given monodromy matrix $W(\lambda)$.

Problem I (existence theorems) was fully and completely solved by V.P. Potapov [6] and in this respect his above mentioned joke is well grounded. From talks with V.P. Potapov we know that for a long time he tried hard to prove the uniqueness of the solution of the formulated inverse problem (Problem II) and his failure to do it, we believe, was caused by his desire to solve the problem with the same degree of generality as the existence problem. Under various additional conditions existence theorems have been deduced in the works [7]–[9], [5], [2]. More complete results have been obtained in the definite case $J = I_n$ (see [2], [8]). As to problem III (the construction of the solution) V.P. Potapov's method is an approximation procedure: the matrix $W(\lambda)$ is approximated by rational matrices–functions $W_N(\lambda)$. The matrices $W_N(\lambda)$ are decomposed into elementary factors. With the help of Helly' theorem the existence of the integral representation (as $n \to \infty$) is proved

$$(1.4) \qquad\qquad W(\lambda) = \int_0^\ell \mathrm{exp}i\lambda J H(t)dt.$$

We remind of the definition of the multiplicative integral in the right–hand side of (1.4):

$$(1.5) \qquad \int_0^\ell \exp\{i\lambda J H(t)dt\} = \lim_{n\to\infty} \exp\{i\lambda \Delta E(t_N)\} \cdots \exp\{i\lambda \Delta E(t_1)\}$$

where $t_0 = 0 < t_1 < \cdots < t_n = \ell$ and

$$\Delta E(t_k) = E(t_k) - E(t_{k-1}) = J \int_{t_{k-1}}^{t_k} H(s)ds.$$

The representation (1.4) is equivalent to the fact that $W(\lambda)$ is a monodromy matrix of the system (1.1). Under substantial additional limitations on the system (1.1) a more effective method of constructing a solution of the inverse problem in question is given in the papers [9] and [10].

Let us note that the classical inverse problems (Sturm–Liouville equations, Dirac type system) were investigated in detail in [3] and [4].

2   Existence theorems

Let us introduce the main notions. By J we shall denote a $n \times n$ matrix with complex entries satisfying the conditions
$$(2.1) \qquad\qquad J = J^* = J^{-1}.$$
The $n \times n$–valued entire function $W(\lambda)$ is called J–contractive in the upper half plane if

$$(2.2) \qquad\qquad W(\lambda)JW(\lambda)^* \leq J, \qquad \mathrm{Im}\,\lambda > 0,$$

and it is called J–unitary on the real axis if

$$(2.3) \qquad\qquad W(\lambda)JW(\lambda)^* = J, \qquad \mathrm{Im}\,\lambda = 0.$$

V.P. Potapov proved the following fundamental theorem connected with Problem I.

Theorem 2.1 *[6] Let $W(\lambda)$ be a J–contractive entire function which is J–unitary on the real line. Then it admits the representation*

(2.4) $$W(\lambda) = \int_0^{\overset{\ell}{\curvearrowleft}} \exp\{i\lambda J H(t) dt\}$$

where $H(t)$ is a summable nonnegative matrix function satisfying the normalizing condition

(2.5) $$\mathrm{sp}\{JH(t)\} = 1.$$

We note that V.P. Potapov obtained a multiplicative decomposition for general meromorphic functions J–contractive in the open upper half–plane.

Let us consider the matrix function

(2.6) $$W(x, \lambda) = \int_0^{\overset{x}{\curvearrowleft}} \exp\{i\lambda J H(t) dt\}.$$

Relation (1.1) then holds. The matrix $W(\lambda) = W(\ell, \lambda)$ is then a monodromy matrix of the system (1.1) with the normalization condition (1.3).

As a consequence of Theorem 2.1 we have:

Corollary 2.2 *(see [6]). Every entire function $W(\lambda)$ which is J–contractive in the upper open half plane and J–unitary on the real line and normalized by the condition (1.3) is a monodromy matrix of a system of the form (1.1).*

Remark 2.3 *Conditions (2.2)–(2.3) are in fact necessary and sufficient conditions to ensure a representation of the form (2.4) (see [6]).*

Thus, Potapov's theorem gives a complete solution of Problem I for systems of the form (1.1) when the monodromy matrix of the system is given. As was pointed out by V.P. Potapov the corresponding inverse problem is in fact a spectral one.

3   Classical examples

We now show that a number of important classical equations can be written in the form (1.1).

Example 3.1 *We consider the system of two equations studied by M.G. Krein [9]:*

(3.1) $$\begin{aligned}\frac{dP_1}{dx} &= i\lambda P_1 - \overline{a(x)}P_2 \\ \frac{dP_2}{dx} &= -a(x)P_1\end{aligned}$$

where $a(x) \in L_1(0, l)$.

Discussion: Setting

$$Q(x) = \begin{pmatrix} 0 & \overline{a(x)} \\ a(x) & 0 \end{pmatrix}, \qquad j = \begin{pmatrix} 1 & 0 \\ 0 & -1 \end{pmatrix}$$

and

$$D = \begin{pmatrix} 1 & 0 \\ 0 & 0 \end{pmatrix}, \qquad T_0 = \frac{1}{\sqrt{2}} \begin{pmatrix} 1 & 1 \\ -1 & 1 \end{pmatrix},$$

we obtain the equations

(3.2) $$jQ(x)j = -Q(x)^*, \qquad T_0^* j T_0 = \begin{pmatrix} 0 & 1 \\ 1 & 0 \end{pmatrix} = J.$$

Let $T(x)$ be the matrix function solution of the system

$$\frac{dT(x)}{dx} = -Q(x)T(x), \qquad T(0) = T_0.$$

Thus, $T(x)$ has the form

$$T(x) = \int_0^{\overleftarrow{x}} \exp\{-Q(s)ds\}T_0.$$

Setting

$$y(x) = (y_1(x), y_2(x)) = T(x)^{-1}(P_1(x), P_2(x))$$

the system (3.1) becomes

(3.3) $$\frac{dy}{dx} = i\lambda J H(x)y(x)$$

where we have set

(3.4) $$H(x) = T(x)^* DT(x).$$

Example 3.2 *(The Sturm–Liouville equation) Let us consider the Sturm–Liouville equation*

(3.5) $$-\frac{d^2y}{dx^2} + u(x)y(x) = \lambda y(x)$$

where the potential u is summable on $[0, \ell]$ and real.

Discussion: We introduce the matrix–function

(3.6) $$G(x, \lambda) = \begin{pmatrix} 0 & 1 \\ u(x) - \lambda & 0 \end{pmatrix},$$

and consider the system

(3.7) $$\frac{dW}{dx} = G(x, \lambda)W \qquad W(0, \lambda) = I_2.$$

The connection between the Sturm–Liouville equation and this system is as follows: the functions $W_{11}(x)$ and $W_{12}(x)$ are solutions of equation (3.5) and

$$W_{21}(x, \lambda) = \frac{d}{dx}W_{11}(x, \lambda)$$

$$W_{22}(x, \lambda) = \frac{d}{dx}W_{12}(x, \lambda).$$

Setting

$$(3.8) \qquad P(x) = \begin{pmatrix} 0 & 1 \\ u(x) & 0 \end{pmatrix}, \quad j_1 = i\begin{pmatrix} 0 & -1 \\ 1 & 0 \end{pmatrix}, \quad T_1 = \frac{1}{\sqrt{2}}\begin{pmatrix} i & 1 \\ i & -1 \end{pmatrix},$$

we define the matrix $A(x)$ by the equations

$$\frac{dA}{dx} = P(x)A, \qquad A(0) = T_1,$$

i.e. $A(x) = \int_0^x \exp\{P(x)dx\}T_1$. Then the matrix–function

$$(3.9) \qquad\qquad W(x,\lambda) = A(x)^{-1}W(x,\lambda)T_1$$

is a solution of the system (1.1) where

$$(3.10) \qquad\qquad H(x) = A(x)^* \begin{pmatrix} 1 & 0 \\ 0 & 0 \end{pmatrix} A(x).$$

Example 3.3 *Let us consider the system of Dirac type*

$$(3.11) \qquad\qquad \begin{aligned} \frac{d\varphi}{dx} &= \lambda\psi + a(x)\varphi + b(x)\psi \\ \frac{d\psi}{dx} &= -\lambda\varphi + b(x)\varphi - a(x)\psi \end{aligned}$$

where $a(x)$ and $b(x)$ are real summable functions. In matrix form the above system becomes

$$(3.12) \qquad\qquad \frac{dW}{dx} = i\lambda j_2 W + H_1(x)W$$

where

$$(3.13) \qquad\qquad j_2 = \begin{pmatrix} 0 & i \\ -i & 0 \end{pmatrix}, \quad H_1(x) = \begin{pmatrix} a(x) & b(x) \\ b(x) & -a(x) \end{pmatrix}.$$

Discussion: Let

$$(3.14) \qquad\qquad V(x)W(x,\lambda) = W(x,\lambda)$$

where

$$(3.15) \qquad\qquad V(x) = \int_0^x \exp\{H_1(t)dt\}T_1.$$

Diferentiating (3.14) we obtain, thanks to (3.12) and (3.15), the equality

$$(3.16) \qquad\qquad JH(x) = V(x)^{-1}j_2 V(x).$$

As $j_2 H_1(x) = -H_1(x)^* j_2$, the multiplicative representation (3.15) and the equality $T_1 J T_1^* = j_2$ lead to the relation

$$(3.17) \qquad\qquad V(x)JV(x)^* = j_2.$$

From formulas (3.16) and (3.17) it follows that

$$(3.18) \qquad\qquad H(x) = V(x)^*V(x).$$

4   Uniqueness theorems

Recall that I_k denotes the $k \times k$ identity matrix and let D denote the diagonal matrix

$$(4.1) \qquad\qquad D = \text{diag}\{D_1, D_2, \ldots, D_r\}$$

where
$$(4.2) \qquad\qquad D_i = \alpha_i I_{k_i}$$

with real numbers α_i all distinct and $\sum_i k_i = n$.

Let us introduce the systems of equations

$$(4.3) \qquad\qquad \frac{dY}{dx} = (\lambda D + A(x))Y$$

$$(4.4) \qquad\qquad \frac{dZ}{dx} = (\lambda D + B(x))Z$$

where $0 \le x \le \ell$. In these equations, $A(x)$, $B(x)$, $Y(x, \lambda)$ and $Z(x, \lambda)$ are $\mathbb{C}^{n \times n}$-valued. The monodromy matrices of the systems (4.3) and (4.4) are defined by the equations

$$(4.5) \qquad W_A(\lambda)Y(0, \lambda) = Y(\ell, \lambda), \quad W_B(\lambda)Z(0, \lambda) = Z(\ell, \lambda).$$

We shall use once more the system

$$(4.6) \qquad\qquad \frac{dQ}{dx} = -(\bar{\lambda}D + B(x)^*)Q, \quad 0 \le x \le \ell < \infty.$$

It is easily checked that
$$(4.7) \qquad\qquad Q(\ell)^* = Q(0)^* W_B(\lambda)^{-1}.$$

We shall decompose the matrices $A(x)$ and $B(x)$ into blocks

$$(4.8) \qquad A(x) = (A_{ij}(x)), \qquad\qquad B(x) = (B_{ij}(x)), \qquad 1 \le i, j \le r.$$

along the partition defined by (4.1).

S.L.Leibenzon [5] using G.Borg's method [1] proved the uniqueness theorem for the system (4.3) under the condition $k_i = 1$ for $1 \le i \le n$. To remove this condition we modify this method to some extent.

Theorem 4.1 *Let the following conditions be fulfilled:*

1. *The matrix–functions A and B have entries in $L_2(0, \ell)$*

2. *It holds that*
$$(4.9) \qquad\qquad A_{ii}(x) = B_{ii}(x) = 0, \qquad 1 \le i \le r.$$

3. *The matrix $W_B(\lambda)W_A(\lambda)^{-1}$ is block–diagonal.*

Then the equality

(4.10) $$A(x) = B(x)$$

holds.

Proof: We set as in [5]

(4.11) $$R(x, \lambda) = Y(x, \lambda)Q(x, \lambda)^*.$$

From (1.3) and (1.6) we obtain the equality

(4.12) $$\frac{dR}{dx} = \lambda(DR - RD) + A(x)R - RB(x).$$

Taking into account (4.5), (4.7), we deduce from (4.3) and (4.6) the relation

(4.13) $$\int_0^\ell Q(x, \lambda)^*(A(x) - B(x))Y(x, \lambda)dx = Q(\ell, \lambda)(I_n - W_B(\lambda)W_A(\lambda)^{-1})Y(\ell, \lambda).$$

Let us denote by $\mathcal{L}$ the space of $n \times n$ matrix–valued functions with inner product

(4.14) $$(R_1, R_2) = \int_0^\ell \mathrm{Sp}\{R_1(x)R_2(x)^*\}dx$$

Since $\mathrm{Sp}\ (R_1 R_2^*) = \mathrm{Sp}\ (R_2^* R_1)$, it follows from the equation (4.13) that

(4.15) $$(R, C) = \mathrm{Sp}\ \{(I_n - W_B(\lambda)W_A(\lambda)^{-1})R(\ell, \lambda)\}$$

where

(4.16) $$C(x) = A(x)^* - B(x)^*.$$

We denote by $\mathcal{L}_1$ the subspace of block–diagonal matrix–functions, the dimension of the blocks being acording to the decomposition (4.1). The subspace $\mathcal{L}_2 = \mathcal{L} \ominus \mathcal{L}_1$ consists of matrix–functions with diagonal blocks equal to zero.

Let P_k be the orthogonal projection from $\mathcal{L}$ onto $\mathcal{L}_k$. Setting $R_k = P_k R$ we rewrite the system (4.12) in the form

(4.17) $$\frac{dR_1}{dx} = P_1(A(x)R_2 - R_2 B(x))$$

(4.18) $$\frac{dR_2}{dx} = \lambda K R_2 + (A(x)R_1 - R_1 B(x)) + P_2(A(x)R_2 - R_2 B(x))$$

where the operators K acts on $\mathcal{L}_2$ and is defined by the formula

(4.19) $$K R_2 = DR_2 - R_2 D.$$

We suppose furthermore that

(4.20) $$R_1(\ell) = 0.$$

Then from (4.17) we have the following equality

(4.21) $$R_1 = T_1 R_2 = -\int_x^\ell P_1(A(u)R_2(u) - R_2(u)B(u))du.$$

Let us now introduce the operators S and T on $\mathcal{L}_2$ by

(4.22) $\qquad SR_2 = \frac{1}{2}\left(\int_0^x R_2(u)du - \int_x^\ell R_2(u)du\right)$

(4.23) $\qquad TR_2 = S\{A(x)(T_1R_2) - (T_1R_2)B(x) + P_2(A(x)R_2 - R_2B(x))\}.$

Then the system (4.18) can be transformed into the form

(4.24) $\qquad\qquad R_2(x,\lambda) = F_R(\lambda) + \lambda SKR_2 + TR_2$

where

(4.25) $\qquad\qquad F_R(\lambda) = \dfrac{(R_2(0,\lambda) + R_2(\ell,\lambda))}{2}.$

It follows from (4.15), (4.20) and condition III of the theorem that

(4.26) $\qquad\qquad (R_2(x,\lambda), C(x)) = 0.$

By differentiating both parts of (4.26) with respect to λ we obtain

(4.27) $\qquad\qquad (\dfrac{\partial^p R_2(x,\lambda)}{\partial\lambda^p}, C(x)) = 0, \qquad p = 0,1,2,\ldots$

Setting

$$L(\lambda)Y = Y - \lambda SKY - TY$$

for $Y \in L_2(0,\ell)$ we deduce from (4.24) that

(4.28) $\qquad\qquad L(\lambda)\dfrac{\partial^p R_2}{\partial\lambda^p} = \dfrac{\partial^p F_R(\lambda)}{\partial\lambda^p} + SK\dfrac{\partial^{p-1} R_2}{\partial\lambda^{p-1}}; \qquad p = 1,2,\ldots$

Let $Y_0, Y_1, \ldots, Y_m$ be now a system of eigenvectors and generalized eigenvectors of the pencil $L(\lambda)$ at the point λ_j. If $Y_{-1} = 0$ the relations

(4.29) $\qquad\qquad L(\lambda_j)Y_p = SKY_{p-1}, \qquad p = 0,1,\ldots m$

hold.

Let us choose the initial conditions in such a way that $R_2(x,\lambda_j) = Y_0(x)$. From (4.28) and (4.29) the equality

$$Y_1(x) = \frac{\partial R_2}{\partial\lambda}|_{\lambda=\lambda_j} + V_1(x,\lambda_j)$$

follows. Here, $V_1(x,\lambda)$ is a solution of the system (4.24). It can be proved in the same manner that for any p $(1 \le p \le m)$ there are solutions $V_{p,1}(x,\lambda), V_{p,2}(x,\lambda),\ldots,V_{p,p}(x,\lambda)$ of the system (3.24) such that

(4.30) $\qquad\qquad Y_p(x) = \left(\dfrac{\partial^p R_2}{\partial\lambda^p} + \dfrac{\partial^{p-1}V_{p,1}}{\partial\lambda^{p-1}} + \cdots + V_{p,p}\right)|_{\lambda=\lambda_j}.$

It follows from (4.27) and (4.30) that the system of eigenvectors and generalized eigenvectors of the pencil $L(\lambda)$ is orthogonal to $C(x) = A(x)^* - B(x)^*$.

Let us note thast the operator iSK is selfadjoint, complete and belongs to the class γ_p and the operator T is compact. Thus, the pencil $L(\lambda)$ satisfies the conditions of Keldysh's theorem. The eigenvectors and generalized eigenvectors of the pencil form a complete system in L_2. From the orthogonality of this system to $C(x)$ the assertion which we are proving follows.

Let us suppose that $H(x)$ in the system (1.1) admits the representation

$$(4.31) \qquad\qquad JH(x) = V_1(x)DV_1(x)^{-1}$$

where the matrix D is determined by (4.1)–(4.2). It is easy to see that

$$(4.32) \qquad\qquad JH(x) = V(x)DV(x)^{-1}$$

where
$$(4.33) \qquad\qquad V(x) = V_1(x)U(x)$$

and where $U(x)$ is a block–diagonal matrix function. Supposing that V is differentiable and setting
$$(4.34) \qquad\qquad \mathcal{W}(x, \lambda) = V(x)^{-1}W(x, \lambda)V(0)$$

the system (1.10) becomes

$$(4.35) \qquad\qquad \frac{d\mathcal{W}(x, \lambda)}{dx} - \{\lambda D + A(x)\}\mathcal{W}(x, \lambda)$$

where we have set
$$(4.36) \qquad\qquad A(x) = -V(x)^{-1}\frac{dV}{dx}.$$

As is known, by choosing adequately $U(x)$ it is possible to obtain the fullfilment of condition (4.9).

If $J = I_n$ and $r = 2$ the smoothness of the function $H(x)$ is not required for the proof of the existence theorems. In this case a substantial role is played by the notion of one–cell [4] and one–block [5] character of the integral Volterra operator.

The papers [5]–[7] are dedicated to the methods of constructing the system (1.1) from the given monodromy matrix $W(\lambda)$. As well as in classical problems ([8]–[9]) the main part in these methods is played by the transformation operator. The method of reconstructing the system (1.1) from the spectral matrix–function $\tau(u)$ has been considered in the paper [7]. The connection between $W(\lambda)$ and $\tau(u)$ is given in the same paper.

References

[1] G. Borg. Eine Umkherung der Sturm–Liouvillschen Eigenvertaufgabe. *Acta Mathematica*, pages 1–96, 1946.

[2] M.S. Brodskii. *Treugol'nnye i zhrdanovy predstavlenija limeinykh operatorov.* Nauka, Moskow, 1969. English translation: Triangular and Jordan representation of linear operators, American mathematical society, Providence, 1971.

[3] I.M. Gelfand and B.M. Levitan. On the determination of a differential equation from its spectral function. *Izv. Akad. Nauk SSSR ser. math.*, 15:309–360, 1951.

[4] M.G. Kreĭn. Continuous analogues of propositions for polynomials orthogonal on the unit circle. *Dokl. Akad. Nauk. SSSR*, 105:637–640, 1955.

[5] Z.L. Leibenzon. A connection between the inverse problem and the completeness of eigenfunctions. *Dokl. Akad. Nauk SSSR*, 145:519–522, 1962. English translation: Soviet math. Dokl., 3, (1962), 1045–48.

[6] V.P. Potapov. The multiplicative structure of J-contractive matrix–functions. *Trudy Moskow. Mat. Obsc.*, 4, 1955. English translation in: American mathematical society translations, vol. 15, p. 131–243 (1960).

[7] L.A. Sahknovich. On reduction of Volterra operators to the simplest form and on inverse problems. *Izv. Akad. Nauk. USSR ser. Math.*, 21:235–262, 1957. (in russian).

[8] L.A. Sahknovich. Dissipative Volterra operators. *Math. USSR Sbornik*, 5:311–331, 1968.

[9] L.A. Sahknovich. On similarity of operators. *Siberian math. Journal*, 13:604–615, 1972.

[10] L.A. Sahknovich. Factorization problems and operator identities. *Russian mathematical surveys*, 41:1–64, 1986.

Pr. Dobrovolskogo 154 ap. 199
Odessa 270111
Ukraine

Operator Theory: Advances and Applications

Edited by
I. Gohberg, School of Mathematical Sciences, Tel-Aviv University, Ramat Aviv, Israel

This series is devoted to the publication of current research in operator theory, with particular emphasis on applications to classical analysis and the theory of integral equations, as well as to numerical analysis, mathematical physics and mathematical methods in electrical engineering.